SAVING BULL CREEK

SAVING BULL CREEK

By Loring Bullard and Robert Kipfer

Foreword by Mike Kromrey

Published by Backyard Naturalist Press

Saving Bull Creek
Loring Bullard
Published by Backyard Naturalist Press
Springfield, MO 65804

Book Design by Randy Clute

Text Editing by Rachel E. Kersey

Cover Art by Jane Troup, Janetroup.com

Publisher's Cataloging-in-Publication
Names: Bullard, Loring, author. | Kipfer, Robert, author. | Kromrey, Mike, writer of foreword.
Title: Saving Bull Creek / by Loring Bullard and Robert Kipfer ; foreword by Mike Kromrey.
Description: Springfield, MO : Backyard Naturalist Press, [2026] | Includes bibliographical references and index.
Identifiers: LCCN: 2026903399 | ISBN: 9798994772126
Subjects: LCSH: Watersheds--Missouri--Greene County--History. | Rivers--Missouri--Greene County--History. | Bull Creek (Mo.)--Environmental conditions. | BISAC: SCIENCE / Natural History. | SCIENCE / Environmental Science. | SCIENCE / Life Sciences / General.
Classification: LCC: GB991.M8 B85 2026 | DDC: 551.480977878--dc23

2026

CONTENTS

FOREWORD

Let's begin this wonderful book with a question to you: *what is your river story?* If you picked this book up, you most likely have one—in fact, almost everybody has a river story. Take a moment to close your eyes and think about a memory or two from the river. What did it smell like? What did it sound like? Was anybody with you and how did you feel?

In preparation for this book, I began asking friends the same question: *W hat is your river story?* The very first person to whom I posed this question was quite taken aback—jolted almost. When she realized the question was serious, her eyes became watery, and you could almost see the memories flooding back. The first and most crystalized memory she recounted was of her first visit to Bull Creek, nearly fifty years ago: "I remember sitting down in the cold water. The memory and water are so clear that I can almost still see the individual pebbles on the bottom. The cool water brought a great physical awareness, the nibbles of the minnows, and the warmth of the sun on my skin. But it also brought another awareness I don't think I've experienced so intensely before or since, which was a tremendous feeling of connection and contentedness."

Another friend shared that coming from central Missouri to the Ozarks was a revelation. "The clean, clear, cool, flowing water with an abundance of different fish and other creatures was astounding to me. I fell in love." Bull Creek is the confluence of tributaries, springs, hills, friendships, learning, history, stories, and loves; it has been centrally important to friendships and kinships in my own life.

Bob and Barbara (Barb) Kipfer felt the call of Bull Creek and eventually graduated from family canoe trips to a full-blown, encore career of full-time stewardship and learning. After purchasing the Bull Mills property near Red Bridge Road, their curiosity and land ethic propelled them to seek out the stories of the place and guidance in their stewardship. They began by contacting professionals with the Missouri Department of Conservation and, soon after, got involved with the Springfield Plateau Chapter of the Missouri Master Naturalists. Their journey includes stream bank stabilization, invasive species removal, glade clearing and prescribed fire, and planting thousands of trees—lasting physical impacts with water quality and habitat benefits.

But perhaps their most long-term impact will be their impacts stemming from their curiosity and enthusiasm for sharing their place and what they've learned. As a student of history and Henry Rowe

Schoolcraft, Bob, on occasion, will wear period clothing and serve as an interpreter and bring Mr. Schoolcraft back to life. Barb's presence on the property makes the invasive weeds shiver in their boots as she patrols with keen eyes and a gunslinger's belt equipped with pruners, gloves, and herbicide.

Upon writing this, we have just completed our 15th annual field trip for the Wonders of the Ozarks Learning Facility (WOLF) fifth graders. For years, Bob and Barb have dedicated countless hours volunteering at WOLF, helping to teach and share their curiosity. The field trip occurs near Halloween, and after several interactive stations of Ozarks-centric experiential learning, they congregate in the Bull Mills Cemetery. Surrounded by primitive headstones, Bob's booming voice echoes in the valley, sharing "real life" ghost stories of the people connected to the place where we are standing. He has pieced these stories together through the years and brings to life the voices of the hills.

Loring Bullard also fell hard for Bull Creek. In his younger days, the call of whitewater and wild rivers beckoned him from all corners of North America. Every form of stream, small and large, captivated him from a young age, and became the through line of his career and life. His vision led to the success of the Watershed Committee of the Ozarks and creation of the Watershed Center, now fostering generations that understand and care for the Ozarks and our water workings. His mentorship has been invaluable to me. His ethos of creating and nurturing partnerships seeped into the DNA of Watershed and that is where most of the successes and progress towards "clean water for life," are rooted. Throughout this journey, Bull Creek continues to beckon him back.

Rivers run on, and through it all, as Norman Maclean best describes. The raindrops in the hills are timeless, and the bedrock of Bull Creek is the basement of time. The flow conjures voices from the past, the future, and within us. Are you "haunted by waters, " too? *Saving Bull Creek* provides a snapshot in time for the voices of our beloved Bull Creek to speak, and helps you continue to write your own river story.

Mike Kromrey, Executive Director
Watershed Committee of the Ozarks

INTRODUCTION

Why write a book about Bull Creek? After all, there are many beautiful streams in the Ozarks, and some of them have had great books written about them. This is especially true for those streams with national recognition—such as the Current, Jacks Fork, Eleven Point and Buffalo—which have a degree of protection as federally managed lands. Like most Ozark streams, Bull Creek does not have any special protection. And it is certainly not unique in being a high-quality Ozark stream. But it *is* unique in being a high-quality stream located in proximity to the largest metropolitan area in the Ozarks.

That nearness lends itself to the old "two-edged sword" metaphor. On one hand, proximity to large population centers makes it easy for urban and suburban folks to access Bull Creek. Hopefully, these folks will appreciate and respect the creek's natural wonders and be careful not to mar or pollute it. However, the surrounding, rapidly growing communities are at the same time pressing outward into Bull Creek's watershed. Without thoughtful management, this expanding development could slowly but surely degrade the quality of the creek and its watershed.

In this book, we hope to prove that Bull Creek is worth protecting. We will stress the need to use better practices in land development, farming, and recreation to ensure that its high-quality waters and healthy watershed remain unimpaired for future generations. But mostly, our motivations to write the book arise from the fact that we both have personal and profound connections to Bull Creek.

One of us owns a significant parcel of land right on the creek and is trying to manage that land in a way that will keep Bull Creek healthy. The other is a canoeist and angler, who relishes the experiences offered by a vibrant, healthy creek so close to home. Late in our lives, we both arrived at Bull Creek, but by quite different roads. We begin with our descriptions of those separate journeys. We end with a plea to the reader to help us make sure Bull Creek remains in perpetuity the gem that it is today.

Red Bridge Road

BOB'S ROAD TO BULL CREEK

Arriving at our Bull Creek property from Red Bridge Road, I turn through a gate and pass under old, rusting metal towers. These carried the powerlines that brought electricity northward from the 1914 Powersite Dam on the White River. Continuing down the lane, I travel the only remaining track of the original 1833 Mail Trace Road between Springfield and Forsyth, lined by tall trees on both sides and barely ten feet wide. Both landmarks bear mute testimony to the significant "improvements" that eventually reached the Bull Creek country. Change came slowly, however, to these wooded hills. The Mail Trace started off like most roads in this rugged region—rough, rutted, and rocky. Bridges were nearly nonexistent in the early days, with most roads crossing creeks at fords. In 1915, a bridge was finally built across Bull Creek just south of our property, lending its name to what would become Red Bridge Road. This single-lane, three-span, pony truss bridge was built by the Canton Bridge Company of Canton, Ohio. Although it is eligible for the National Register of Historic Places, its fate is sealed, as it is scheduled to be replaced by a stronger, concrete, two-lane bridge. Unfortunately for us, this will undoubtedly encourage faster traffic and larger trucks on the narrow chip and seal road. The changes continue, not all of them pleasing or desirable.

My wife, Barb, and I stumbled upon our Bull Creek property somewhat by accident. At the time, I was a practicing gastroenterologist and Barb was a hospital nurse. Immersed in our medical practices and raising a family, our fun on weekends was mostly confined to canoeing, fishing, and sailing on local streams and lakes. By 1995, our son and daughter had both left to start their own careers, providing us with some additional free time.

Barb and Bob Kipfer at their Bull Creek property

Red Bridge by Steve Spencer

The "Shelton" cabin on Kipfer's property

With an empty nest, Barb and I began looking for a place to get away and relax in the country, relatively close to our home in Springfield. We made an offer on a property east of Sparta, near Swan Creek, but while that purchase was pending, our realtor suggested a place on Bull Creek. That sounded promising, especially since I had heard talk around the hospital about Dr. Shelton's "beautiful land on Bull Creek." On a whim, we drove down to see it in December 1995. We arrived after sunset, with just enough light left to see the rustic road and steep hills rising above the valley. We forded the creek and around a curve came upon a small, A-frame house. A warm glow from the fireplace inside greeted us. We were hooked. As time passed, we gradually added on pieces of

Barb and Bob Kipfer at Bull Creek

adjoining woodlands and stream banks. Dr. Shelton sold us his cabin in a hollow, the site of several cabins over the past century.

Until buying that land on Bull Creek, our view of nature had mostly been over the bow of a canoe or kayak. I grew up in Wichita, Kansas, where my father worked as an engineer with Boeing. He took me fishing on the nearby Arkansas River, where I pulled catfish and carp from the roiling, muddy stream. I learned how to canoe on its sluggish waters. These were pleasant experiences, but like most people, I viewed nature as something spelled with a small "n,"—a nice backdrop for fun and relaxation, but not much more. When Barb and I first arrived at Bull Creek, we still thought that nature was something to be admired from a distance.

But that soon changed. We began enjoying long walks along the creek or hiking through the woods. I started fishing for smallmouth bass in the creek's deep, crystal-clear swimming holes. Our appreciation for Bull Creek, and the natural world in general, began to grow. Early on, we also saw what Mother Nature can unleash and how vulnerable our property was to her whims. Heavy rains causing the great Mississippi Flood of 1993 also deluged southwest Missouri. Raging waters in Bull Creek caused severe erosion of the streambank across from our house, excavating a large fishhook curve that threatened to carve a new channel across the field. We looked for help from the Missouri Department of Conservation. Their stream professionals proposed a bank stabilization project that would include planting trees in the barren riparian area.

The new "American Gothic." Bob and Barb at a prescribed burn

At the time, I didn't even know what they meant by "riparian area." Our Conservation Department agent, Larry Martien, explained that it is the vegetated zone along the stream, an area very important to its health. The plan called for planting two-hundred trees and installing riprap along heavily eroding banks. Larry patiently took us through the cost-share process. The next year, we watched from our deck as dump trucks unloaded enough big rocks to make our streambank resemble Table Rock Dam. With the help of a borrowed tractor, we found ourselves planting 2,000 additional trees.

Within a few years, we found that our "getaway time" began to resemble real work. As part of a glade restoration project, we spent weekends cutting cedar trees on a steep, south-facing slope and hauling them into piles. The next year, we performed our first prescribed burn, assisted by more new friends from the Conservation Department. By now, word around the Department headquarters was "we have a live one hooked," and other agents began proposing new projects like timber stand improvement and warm season grass plots.

It was a lot of work, but we gradually learned why these practices were worth the time and effort. As the species of grasses and forbs increased, so did the numbers of birds and other wildlife. Watching deer and turkey, we learned the importance of acorn mast and other natural food sources. Our forester explained timber stand improvement, advising us to remove certain trees to encourage the growth of young white oaks, which in turn would provide acorns for the deer, turkey and even bears. We girdled trees with chainsaws, cutting the bark around the base to kill them, creating

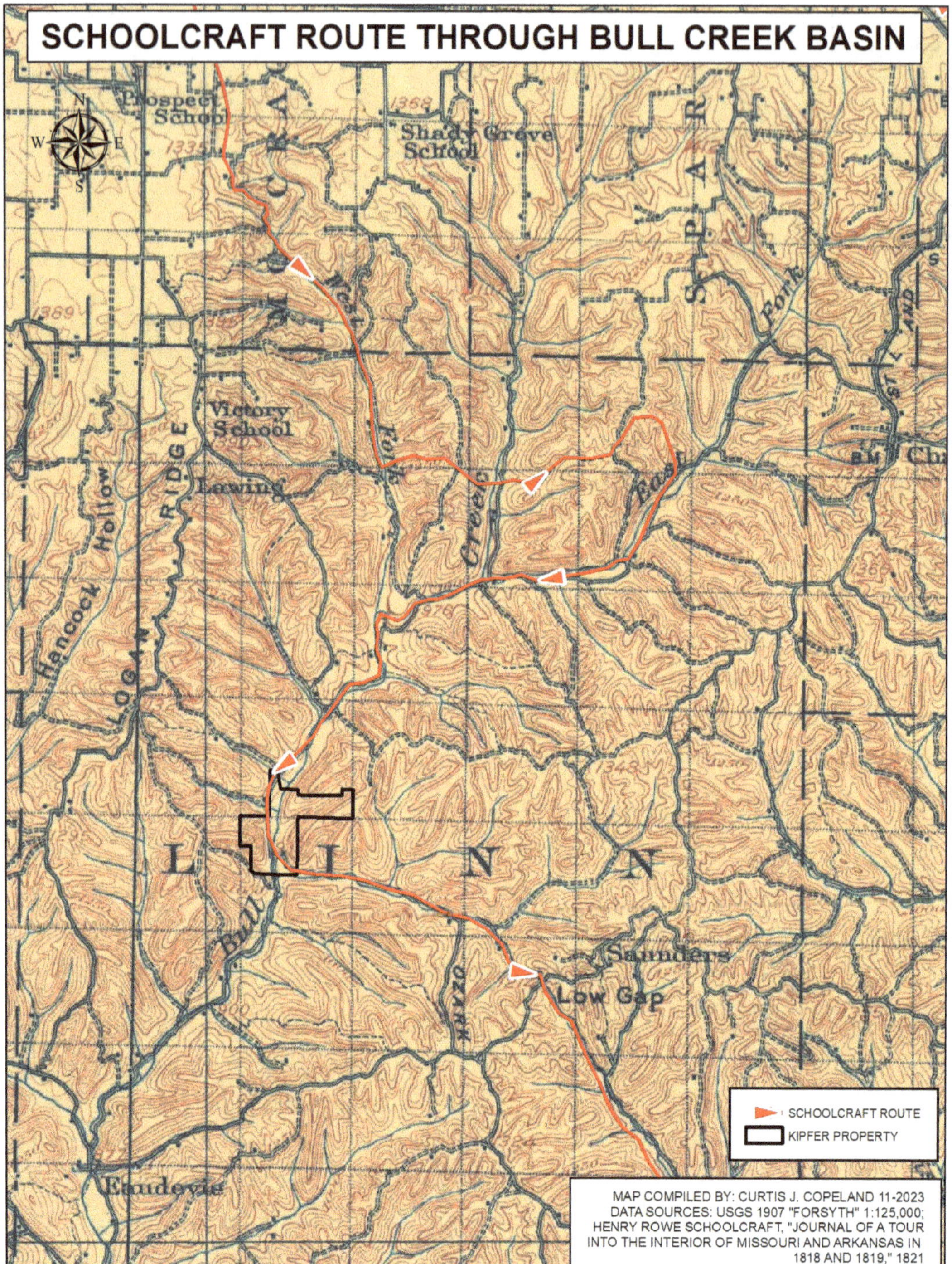
SCHOOLCRAFT ROUTE THROUGH BULL CREEK BASIN
Prospect School
Shady Grove School
Victory School
Lawing
West Fork
Creek
East
Fork
Hancock Hollow
LOGAN RIDGE
Bull
L I N N
Saunders
Low Gap
OZARK
Eaudevie
SCHOOLCRAFT ROUTE
KIPFER PROPERTY
MAP COMPILED BY: CURTIS J. COPELAND 11-2023
DATA SOURCES: USGS 1907 "FORSYTH" 1:125,000;
HENRY ROWE SCHOOLCRAFT, "JOURNAL OF A TOUR INTO THE INTERIOR OF MISSOURI AND ARKANSAS IN 1818 AND 1819," 1821

Kipfer's barn near Peckout Hollow, where Schoolcraft left the Bull Creek Valley

harvestable timber and light-filled openings on the forest floor. New oaks and sassafras plants sprouted in this open space. Sassafras, we learned, hosts larvae of the beautiful black and blue spicebush swallowtail butterfly. Butterflies became a new passion.

Over time, we learned that our property on Bull Creek had historic as well as biological importance. I had always been interested in the Ozark travels of Henry Rowe Schoolcraft, as related in Milton Rafferty's book *Rude Pursuits and Rugged Peaks,* and had done Schoolcraft reenactments for several years. Now I was excited to discover that the path Schoolcraft followed on January 6, 1819 ran right through our property on Bull Creek! We also learned more about the history of the cemetery and the water-powered mill on our property. After connecting with the Missouri State University Center for Archaeological Research, we soon had Stephen Dyle exploring our 1840s cabin, post office and mill sites. He documented the cabin structure, but also researched and described the early commerce of the area called Bull Mills.

With further research, we learned that the tree-lined road under the north bluff line of the creek was a section of the 1830s Ozark Mail Trace that ran from Springfield to the White River,

Barb and Bob teaching the WOLF School kids at Bull Creek

ending near the future site of Forsyth. Headstones in the Cobb-Keaton Cemetery along the road on our property slowly gave up their stories of the people who had settled in the valley over the last two centuries. From the accounts collected by county historical societies, we developed a greater appreciation for the lives and hardships of those pioneers.

We also began to understand the changes that civilization brought to the landscape, and how those changes affect Bull Creek. The construction of the mill on our land in 1839 encouraged an influx of settlers and stimulated commerce. Trees were felled for construction of buildings and bottomlands were broken by the plow. Farming right up to the creek led to severe bank erosion and large gravel bars. Expanding railroads created a demand for wooden ties, resulting in extensive logging on steep hillsides. Later, paved roads encouraged even more settlers, who migrated daily to the cities for jobs or school. A rising interest in outdoor recreation meant more dirt bikes and ATVs racing over thin, rocky soil, creating more erosion. Houses proliferated on ridgetops, with grass lawns mowed weekly by image-conscious suburbanites.

All these changes suggested to Barb and me that we needed to do our part to educate people about land stewardship—leaving the land in better shape than you found it. After retirement, I joined the Springfield Plateau Master Naturalists, which had just formed. Barb retired two years later and together we began new full-time careers as novice naturalists. From membership in the Master Naturalists and other organizations, we learned how to teach others about the wonders of nature.

On our property, we discovered a broad range of plants and insects, many of which only became visible when we learned to really open our eyes, and our minds. My cargo pants pockets soon filled with bug boxes and magnifiers, my macro camera a constant companion. New discoveries led to new friends as we bonded with experts, gifted amateurs, and enthusiastic neophytes. By this time, the Internet was in full bloom with *Wikipedia, BugGuide* and *iNaturalist* providing research rabbit holes to plunge down with each new find. We started a Master Naturalist Blog (Springfieldmn.blogspot.com) in 2009 and it rapidly became a full-time passion, with over 1,400 posts and counting. A chance encounter with a bear under our deck led to working with the Missouri Department of Conservation (MDC) in trapping, radio collaring and tracking bear populations in the area. Seeing feral hogs running trails on our property, we learned how to identify hog signs, and soon MDC had traps on our property, allowing the department to catch and kill a few dozen of the invasive swine.

In 2010, we began hosting field trips for the new Springfield Public Schools choice program, the Wonders of the Ozarks Learning Facility (WOLF School). In this program, students get the standard fifth grade curriculum "taught through the lens of nature." Our weekly classroom programs and field trips ranged from water conservation to invasive species followed by hands-on, small group sessions featuring some of the fascinating productions of nature. Our attic soon filled up with insect specimens, plant galls and other interesting finds.

In our efforts to expand opportunities for nature education, we invited students and researchers to use our Bull Creek property as a research center and field station. Dr. Deb Finn and stream ecology students from Missouri State University conducted research on life in the hyporheic zone, the water that flows under and through gravel bars. We also hosted ornithologists studying the songs of Carolina wrens. Grad students became a regular part of the Bull Mills community. Few eighty-year-old retirees get to hang out with and learn from young graduate students.

Wonders of nature continue to excite and inspire us. One researcher, Nathan Dorff, found a new species of frog-biting midge (*Corethrella sp.).* These are tiny flies with wings less than one tenth of an inch long. The females are obligate external parasites that feed on the blood of tree frogs. Nathan captured them by playing the recorded song of a female tree frog. If pollution or damage to the riparian area had eliminated this midge, we would have never known of its existence. Would that matter? In our opinion, everyone gets a vote. If enough vote, "yes" with their actions to protect and preserve nature, the species may survive. The midge is now officially recognized by the name *Corethrella kipferi.* How cool is that?

Bull Creek runs through the middle of our property. It is still a healthy stream, listed in the state's water quality standards as a biological reference stream, against which the health of other streams can be measured. But increasing urbanization and agricultural runoff in its watershed mean that protecting it will be an ongoing challenge. To help with that, we have established a stream conservation easement with the Missouri Department of Conservation, protecting our river corridor from development. We have also created an irrevocable trust with Missouri State University, which will eventually take over the care and management of our property. Trustees will oversee the agreement, with a provision that the land cannot be sold for fifty years.

Bull Creek has taught us important lessons. As more of us crowd into the watershed and replace forest land with manicured lawns, the additional runoff increases the frequency of flooding. We learned this lesson the hard way when we left our truck parked in a bottomland field along Bull Creek. A twelve-hour rainstorm flooded the valley, filling our truck to the cup-holders. We saw where huge logs and six-hundred-pound hay bales had floated down the valley fields. Perhaps, we thought, this was the creek's way of saying, "Don't even think about building in my valley!"

Bull Creek and its beautiful valley have many more lessons to teach us about history, natural processes, and deep ecology. We have only to keep our senses tuned to the amazing sights, sounds and rhythms of nature. We look forward to the many ways that the creek and our property at the forgotten site of Bull Mills can extend that appreciation for nature to larger audiences and future generations. That, we hope, will be the legacy of our cherished paradise in the valley of Bull Creek.

Loring's Road to Bull Creek

"They paved paradise and they put up a parking lot," Joni Mitchell sang in 1971, the year I started college. At the time, I had no idea what paradise might be. But after a few years in the school's Outing Club, floating and camping on beautiful Ozark streams, I knew I had found the closest thing to paradise that I could imagine. And as a new floating enthusiast, I became aware that over much of the country, paradises like these were slipping away—more and more of them being dammed, or polluted, or otherwise altered and degraded.

The other line, "You don't know what you've got 'til it's gone," also hit a nerve with me. Through my experiences with the Outing Club, I had communed with enough Ozark streams to know full well what I had. Within a few hours' drive were some of the most spectacular and beautiful places I had ever been. I knew these rivers were fun—and scenic, and seemingly alive, although the idea of a river as a living thing had not yet fully formed in my mind.

By the time I became acquainted with Bull Creek, I had a much better idea of what it meant for a river to be truly alive. To me, this creek fit the description perfectly. It was hard to put into words, though. It was more of a feeling, a vague sense that the river was much more than the sum of its parts. It was also partly an illusion—a naïve notion that pristine places like this would always be there, undiminished in quality, available to me whenever I wanted to go.

A healthy Bull Creek smallmouth bass

It could be said that my "road" to Bull Creek began at the farm of my grandparents when I was in grade school. The road was really a set of rutted tracks splattered with cow patties. Back then, when our family visited my grandparents' farm near Green Ridge, Missouri, I eagerly anticipated setting foot on that dusty lane. It took me around the sagging barn, where cats prowled in dark seclusion, before heading down a gentle slope toward the hedge row. Here, the tracks turned south, paralleling a row of twisted Osage orange trees, with hundreds of softball-sized hedge apples in the grass underneath. On some trips, I gathered a bagful of these knobby green fruits to take back home. My dad said they exuded natural pesticides, and he set them on shelves in our dungeon-like basement to ward off insects and spiders.

Meandering southward along the hedge row, the lane finally reached the far corner of the pasture, where thick, deeply weathered posts were bound and tightened with twisted wire. It was easy to duck under the fence there and enter the scrubby forest on the other side, a stand of scrawny, thickly set trees festooned with wildly looping grapevines. As I remember it, the interior of this little patch of forest seemed gloomy and mysterious in the filtered afternoon light—a dream-like forest cathedral.

Not far into the woods, past blackberry tangles and sumac thickets in sunlit openings, I came to my secret abode—the creek—a small rivulet, really, but always flowing; its water a little murky, yet clear enough that I could see crawdads shuffling across the muddy bottom. I followed the creek farther into the forest, to the most secret and magnificent of all my places—the railroad trestle. Where it spanned the creek, sunlight poured in. Shadows from ties and timbers threw a mellow, dappled light on the vegetation below, making me feel like I was inside the terrarium I built at home.

I would sit for a while under the trestle, inhaling the tinge of creosote oozing black and tarry from the timbers, waiting for the next train to rumble by. Trains were a common and, back then, welcome occurrence. I loved trains, especially the visceral sensations as they chugged overhead, spewing black diesel smoke, their hot, oily breaths rolling over me. If I happened to see the engineer leaning out his window, I would always do the air tug and he would usually wave and blow his bone-jarring horn. Railroaders seemed kid-friendly back then.

Between trains, I remained quiet and alert, occupied with observing wild nature in all its glory. Sometimes there were wading birds to spy, rummaging in the shallows, or red-winged blackbirds rasping from the cattails. Occasionally I would see woodpeckers or crows, or even a mangy coyote

Loring in a Bull Creek swimmin' hole

loping across the nearby pasture. The secret place seemed roomy and remote back then. Only later did I realize that it was miniscule, in the grand scheme of things—and isolated, separated from similar wooded patches by wide expanses of farm fields. Habitats were highly fragmented and heavily stressed, even back then.

In fact, that little spot by the creek was marred in ways that I didn't fully realize until much later in life. Invasive species were no doubt crowding in and taking over after many decades of heavy-handed plowing, grading, and logging of the surrounding fields and woods. There was obvious erosion along the creek banks, with dark soil scoured from beneath tangled, overhanging orange roots, and a deep slurry of silt on the stream bottom deposited by runoff from the surrounding farms. But back then, I took little notice of such things. To me, the little sunlit spot along that nameless creek seemed like the very crux of untouched wildness. I always looked forward to the lull after lunch and a leisurely walk down to the creek—to my secret spot.

Those experiences at the creek—following that faint set of tracks to its exquisite, tangled, microcosmic destination—would finally lead me to Bull Creek, a life journey like most marked by memorable turning points—personal misfortunes, in some cases, but also mind opening events. When I turned ten, for example, my parents presented me with a microscope. It was inexpensive, but it was reassuringly solid and heavy, with a barrel operating smoothly on greased gears. It had three magnification settings, as I recall, from ten to three-hundred power. That ingenious device, wielding its magic on the rickety tabletop in my bedroom, opened a whole new world for me.

I began looking at anything and everything under magnification—pennies, grains of salt, feathers. But I quickly learned where the real excitement could be found—peeking into the secret lives of the multitudes of tiny creatures swimming or drifting or jerking erratically through ponds and puddles—rotifers, amoebas, water fleas, hydras—things so incredible and strange that they seemed like alien life forms. What an amazing diversity of life is to be found in every spoonful of water from a lazy creek or muddy slough!

While that tenth birthday began with the promise of scientific enlightenment, it ended in family tragedy, as my grandfather died that very day. After that, trips to the farm became fewer and farther between and before much longer, stopped altogether. I missed my grandfather, of course, but it also hurt that I no longer had access to the creek, the trestle, and the joy of immersion in that wild little corner of Pettis County. I missed it so much that when I was about fourteen, I wrote a story about it called "The Yellow Cat," pecked out on my father's clunky old typewriter. In fact, I still have the story, aged and brittle, kept in a box for over fifty years. I was proud of my writing, obviously, and what I saw as my clever use of irony.

In the story, I was the father of two young children. We were cruising in the family car on a blacktop road west of Green Ridge. I had been telling my kids about the secret spot; how much they would love it there, just as I had loved it as a kid. I told them about my adventures at the farm; about following the rough tracks to the creek; about rummaging through abandoned sheds and peeking into the rusted innards of discarded farm equipment, rising like frozen mechanical insects from the weeds. In those days I was like an old yellow tomcat, I told them; like one they might see prowling around the barn; on his own, carefree, curious about everything, poking his nose and paws into every nook and cranny.

I could sense the interest and excitement building in my kids as we neared the farm, ready to begin their own adventures. But as we pulled up to the spot, I stared out the window in disbelief. Instead of verdant green pastures and ragged hedge rows and the familiar old farmhouse, I looked upon a scene of utter destruction. The entire area had been graded into muddy oblivion, with bladed brown scars of new streets, burning piles of brush and rows of houses under construction—a housing development! On one corner of the desecrated ground sat a big yellow bulldozer—the ironic other yellow cat. I fought back my tears and kept going as my kids questioned me—where is it, daddy? I guess I forgot where it was, I answered.

Years later, the story still triggered that familiar feeling of emptiness, like finding the picture of a long dead loved one. Through the story, I revisited that sense of helplessness in the face of a great injustice—the violation of a secret sanctum. And the story recreated and reinforced that dreadful moment when I realized, and then saw as examples repeated over and over, that travesties like these were happening all the time and everywhere, not just in my make-believe story.

I saw it when I lived in Kansas City, as wooded lots where we played hide-and-seek and war games were systematically bulldozed, stripped of their brush and its hidden warren of trails and forts. It happened later in my small hometown, when bulldozers clearing land along the old slough, a remnant of the channelized South Grand River, forever erased the places where we camped and seined for bullheads and built fires to celebrate life in the wild. And it happened many years later, when I saw so many of my favorite out-of-the way spots in the Ozarks cleared, graded, or developed.

Many years later, I had another turning point, another episode of bitter disappointment—this time, at my new sacred place—Bull Creek. During the installation of a new bridge over the creek, the construction crew dug a straight, deeply cut channel right down the middle of the streambed, with ten-foot-high mounds of gravel piled on either side. Downstream of the work site, a layer of mushy silt blanketed the normally clean gravel bottom. When I first laid eyes on this mess, my heart sank. My fabulous Bull Creek had been destroyed!

It was a horrible feeling, like the one in the story—the same one that inspired me to spend my life trying to protect places like this. The sense of loss, the wounding, was palpable. Who could do this to such a magnificent place? And Bull Creek is truly magnificent; a magical place, in fact, standing out even among the many beautiful streams carving their ways through the verdant hills of

Beth Bullard cooling off in Bull Creek

the Ozarks. Its waters are normally crystal clear, as clear as any stream out there. For much of its length, it has the protection of heavily wooded banks on both sides. But in its lower reaches, it careens dangerously close to "civilization." To urban areas and development and super-highways.

That proximity makes Bull Creek even more attractive to city dwellers like me, who cherish the idea of natural delights so accessible and close to home. One hesitates to call attention to his or her favorite places, for fear of them being overrun. Some of our state's most scenic rivers are being overwhelmed at times—loved to death. On the other hand, the people who know and care about such places are the ones most willing to invest the time and energy in saving them. To those people, we offer this book about our favorite place, in the hopes that you will join us in caring about it. Even more, we hope you will work to protect it—so that this paradise will never slip away.

Opposite: Grading work during Bull Creek bridge construction

CHAPTER 1

A HEALTHY WATERSHED

Bull Creek is a beautiful Ozark stream, a small tributary of the mighty White River of Arkansas and Missouri. The waters of Bull Creek run crystal clear, but what is more important is the great diversity of life found in those waters. Biologic diversity is a good way to measure a stream's health. The more kinds of life that are found in the stream—the more species of animals and plants—the better the chances that very little pollution is coming from the watershed. By this measure, Bull Creek is very healthy indeed.

The Bull Creek watershed features rolling grasslands in the north giving way to rugged hills and deep valleys farther south. Farther south yet, near the creek's mouth, lie the rugged White River Hills. Limestone bedrock is close to the surface on these isolated hills or "knobs." With little overlying soil, trees are stunted, if they can grow at all. Open areas, called glades, are covered with sparse grasses like bluestem and dropseed, giving them the local name of "balds."

The maze of steep, winding hollows in lower Bull Creek formed an impediment to exploration and settlement. While traversing the Ozarks in the winter of 1818-1819, Henry Schoolcraft, a visitor from New York, got turned around and lost there. In that confusing terrain, he spent "one half of the day in vain and perplexing endeavors, wandering from one high knoll to another." Today, with the powerful tools of satellite images and global positioning systems, we can see those complex patterns and features laid out right before our eyes. Sitting comfortably at home, from the viewpoint of a satellite cruising hundreds of miles above the Earth, we can survey the entire watershed of Bull Creek, from its headwater reaches south of the town of Ozark, Missouri, to its mouth in the White River, now buried under a man-made reservoir called Lake Taneycomo.

Views of the Bull Creek watershed from space are artistically compelling. Forested headwater drainages reaching northward into grasslands resemble dark leaves spread on a light green tablecloth. The whole watershed, in fact, appears tree-like, or dendritic, with headwater drainages, like branches, becoming progressively smaller toward the tree-top. The tree that is Bull Creek leans

Opposite: Google Earth image of the Bull Creek watershed. U.S. 65 Highway runs down the west side of the watershed.

to the right, as if a western wind is blowing, bending its twisted trunk. Mature trees clothing the hillsides add rough, dark green texture to images from space. This extensive forest cover is an important key to the health of the watershed. Forests add nutrients and oxygen, critical for all life, and act like a protective blanket, keeping sediment and pollution out of the stream.

Viewed from space, land uses and vegetation types in the Bull Creek watershed can easily be distinguished by their shapes, textures, and colors. In the northern part of the watershed, rectangular pasture lands are interspersed with smaller suburban lawns, all straight-edged, smooth, and bright green in spring or summer. Farther south, at the escarpment separating the Springfield Plateau from the Salem Plateau, the land falls away sharply into steep, mostly forested drainages—irregularly shaped, dark, and rough-textured.

From abrupt changes in vegetative cover, apparent on aerial images, it is easy to see where tiny rivulets end their casual moseying across nearly level pastures or suburban yards and drop into the

Bull Creek tributary after a rainstorm

rocky chutes of tiny, forested headwater streams, picking up speed on their steep descent toward Bull Creek. They are quickly joined by similar rivulets from the adjacent hillsides, which mix and tumble over limestone rocks and ledges, joined by other branches of the watershed-tree before coalescing into even bigger branches downstream, all merging and mingling to become a main trunk, a torrent of swirling waters at the lower end of the watershed.

A heavily forested watershed is a key to the clarity of Bull Creek. Leaves absorb the impact of raindrops, preventing erosion at the land surface. Most sediment is removed as runoff is slowed and filtered by a thick blanket of fallen leaves. The result is a creek that is amazingly clear, about as clear as any stream could be. This attribute of Ozark streams confounded Schoolcraft, as the "extreme limpidity" produced, in his mind, a "species of deception." In attempting to ford the North Fork River, where the water looked at most three feet deep, "the horse suddenly plunged in below his depth, and was compelled to swim across, by which our baggage got completely wetted."

Bull Creek's outstanding clarity is due not only to an extensive forest cover, but also to the geologic underpinnings of the watershed. The creek drains southward off the Ozark uplift, a broad dome of carbonate rocks, predominantly limestone and its close cousin, dolomite (or dolostone). As these rocks erode, the harder, imbedded crystals of chert, or flint, are left behind. These angular pieces of chert cover the streambed and pile up in places as gravel bars. Creek water flowing through and over this huge mass of gravel is strained of most of its particulate matter, including the remaining bits of clay or soil.

The bed of Bull Creek, like those of neighboring streams, is composed almost entirely of chert gravel. It is angular, of various sizes, and mostly shades of brown, off-white, rust, and tan. Gravel is a natural component of Ozark streams—a sharp-edged, granular counterpoint to the fluid, dynamic circulation of currents and eddies. Working together, gravel and water shape the channel, especially during floods—partners in a never-ending dance as rivers sashay across their floodplains, pushing against gravel bars on either side.

River gravel in Bull Creek is super-abundant, to the point of sometimes becoming a nuisance. It clogs culverts under bridges and fills in fishing and swimming holes. During floods, it moves in aggressive, submerged slurries, eroding and abrading, ripping out banks, knocking down trees and fences, sometimes burying productive farm fields. County road departments often excavate gravel at low-water bridges, killing two birds with one stone—removing gravel that has filled bridge openings,

Bull Creek has many large gravel bars

while at the same time obtaining gravel for road and building projects. At some point, demand for river gravel as a construction material became high enough to entice commercial mining operations, some of which set up on Bull Creek.

Much of Bull Creek's watershed is karst topography, riddled with fractures and caves and pockmarked with sinkholes. Landscape features like losing streams and sinkholes quickly funnel surface water underground to reappear at springs. In other places, this shallow groundwater oozes upward, unseen, into the gravel of creek beds. Springs make the water of Bull Creek refreshingly cool, even in the heat of summer. They also supply much of the creek's base flow during dry seasons. Springs are thus vital to the ecology of the stream, sustaining year-round aquatic habitats for a unique and diverse set of organisms.

An Ozark spring

Like tree cover, the riparian zone, the ribbon of bottomland vegetation abutting the streambanks, is critical to stream health. Tree roots hold streambank soils in place and overarching shade keeps water cooler in summer, allowing it to hold more oxygen. Roots and logs provide habitat for fish and a variety of other aquatic life. Leaves are a primary food source for many aquatic organisms, and streambank vegetation slows and filters floodwaters, lessening erosion and downstream siltation. Bull Creek is in remarkably good condition precisely because much of the watershed remains forested, and over most of its length, streambanks are still held in place by trees.

In the past, the geography of Bull Creek has provided it with a degree of protection. Thin soils on steep, rocky hillsides are hard to farm, discouraging the tilling of soil. Narrow, steep-sided hollows made access difficult for loggers and miners. Settlers usually looked for flat, level ground to build their cabins. As a result of these factors, most of the Bull Creek watershed has remained covered in forests, rather than converted to fields, pastures and homesites. At the same time, this rugged, forested topography has made the watershed a mecca for wildlife that shuns areas more extensively farmed or developed.

But the watershed's geography also makes it vulnerable to degradation. Just north of the watershed lies the metropolitan area of Springfield, containing the rapidly growing satellite communities of Ozark and Nixa. To the south, the urban zone around Branson is also expanding. These communities put pressure on the Bull Creek watershed from both sides—from land clearing, off-road recreation, cattle grazing, commercial and residential developments, and—potentially—habitat and water quality degradation that often go with these activities. Bull Creek, it seems, is directly in the way of progress—although, what version of progress would involve the loss of a gem like Bull Creek?

CHAPTER 2

THE NATURE OF BULL CREEK

The Bull Creek watershed has been sharply sculpted by flowing water. Deeply carved headwater streams drop off level highlands to the north, sending water southward toward the large White River. The hills and valleys are, for the most part, clothed in forests, while pastures spread on flatter ridgetop divides and in the larger valleys. Bull Creek flows in a serpentine manner down its narrow valley, curving against bluffs in places and piling up gravel bars at almost every turn. The clear water of the stream runs though swift chutes in some places and pauses in others as deep blue pools. This varied topography, largely untouched by the hand of man, produces a scene most pleasing to the human eye. It also produces a variety of habitat types, both aquatic and terrestrial.

Bull Creek did not always look this way. In cooler times, for example, spruce trees might have been growing in the Bull Creek hills where oaks, pines or cedars now stand. Glades, dry upland areas with shallow soils and sparse grasses, might have been more extensive during the drying times after the last Ice Age. The stream valley no doubt looked much different several hundred years ago. Along many, if not most, of the headwater streams were wide, shallow pools of standing water backed up by dams made of sticks and mud. All this was the work of a remarkable natural engineer—the lowly beaver.

Today, it is hard to imagine how profoundly beavers modified prehistoric landscapes. For millennia, tens of millions of these rodents inhabited North America, in suitable habitat as many as fifty animals per square mile. Collectively, they constituted a powerful hydrologic force. By storing water behind thousands of low dams, they controlled floods. Widespread shallow pools of standing water allowed time for fine sediment to settle to the bottom. Building up over the years, this fine sediment, or silt, created rich floodplain soils. We now know that the fertile farmlands of Iowa, Illinois and north Missouri largely owe their existence to the prehistoric work of beavers.

Beavers are considered a keystone species. Their massive public works projects created ecological niches for countless other living things. Grasses, sedges, and wildflowers grew on pond shorelines and

Opposite: Beavers have just started this dam on Bull Creek. It will probably wash away in the next flood.

flanking wet meadows. Rooted aquatic plants like lily pads and arrowroot inhabited clear, shallow water. Sunfish and bass swam in the deeper pools. Muskrats, otters, and minks cohabitated with beavers in the valley, along with bear, elk, and deer. The loss of beavers produced a ripple effect felt across the entire ecosystem. With beaver dams gone, streams began eroding into their floodplains once more. Valuable soil and nutrients were carried downstream. And the myriad plants and animals that depended upon beaver ponds quickly disappeared.

Beaver, Castor canadensis, *a large rodent*

Beavers are sometimes considered pests in developed areas when they cut down desirable trees or back water onto properties. But important watershed services are lost when they are gone. In their absence, other measures must be taken to prevent flooding and erosion problems, such as maintaining healthy vegetated banks along streams. In the Bull Creek watershed, a healthy population of beavers can be counted on to provide myriad benefits for the stream, and for the health of the watershed.

Like beavers, black bears were wiped out, or nearly so, over a hundred years ago. When they were common, they wielded a considerable influence in the watershed. Black bears eat a lot of fruit and disperse seeds over a wide area. Their fecal waste helps to fertilize the soil. As part-time predators, they keep the populations of other animals in check, helping to maintain an ecological balance. Bears are good for the watershed, and watersheds that contain bears can be considered healthy.

Bears were common when Schoolcraft trekked the Ozarks in 1818-19. They were hunted for hides and oil, and bear meat was commonly on the table. Schoolcraft frequently saw bear products at settler's cabins. Hunters usually tracked them down in early winter dens when they were at their fattest. Bear fat was often turned into oil. It did not turn rancid like butter and other oils and had a longer shelf life. It was shipped to Europe for hair dressings and even lit the streetlights in New Orleans. There were plenty of bears in the early days, so lots of oil was produced. Between 1809 and 1811, John Shaw and two companions reported taking 300 bear skins and 800 gallons of bear oil along the White River.

The bear population in Missouri declined rapidly after the first decades of the nineteenth century. By 1894, some reports suggested that Missouri bears had become extinct. However, there were occasional unverified reports of individual bear sightings into the 1930s. Arkansas began restoring its bear population in 1957. By 1968, over 200 bears had been brought into Arkansas from Minnesota at a cost of over $600 apiece.

The Missouri Department of Conservation contemplated a bear re-stocking project but decided against it. "In stacking up the territorial requirements, diversity and abundance of fruiting trees and shrubs needed, as well as range undisturbed by humans, there is little doubt that what Missouri has to offer is, at the very best, highly dubious in supporting anything like a huntable population." As the Arkansas bear population grew, however, their range expanded into Missouri and parts of Oklahoma. Missouri bears proved to be more adaptable than had originally been predicted.

In 2010, the Missouri Department of Conservation began a large-scale study of Missouri's black bear population using modern technologies. Bears were trapped with cage or leg traps, sedated, and fitted with GPS radiotelemetry collars and ear tags. Signals from the collar were used to create maps, which were posted online a week later. This delay protected the bear's real time location. Bear locations could also be tracked using DNA technology. To accomplish this, a 20-inch-high corral made of single-strand barbed wire was used to enclose a pile of bait food. When the bear stepped over the wire, it would leave hair samples on the barbs. Over time, a travel profile for the bear could be plotted by using multiple DNA samples.

The survey suggested that a few native bears had remained in Missouri, with the department announcing, "Our largest population, in south-central Missouri (Webster and Douglas counties), may represent a small remnant of that region's historical population, combined with bears descended

Bob with a radio collared bear

from the Arkansas releases." A statewide estimate in 2012 suggested a population of just under 300 bears. An updated estimate in 2019 determined there were about 540-840 bears in the state, with the population growing about nine percent annually.

Bears now roam over the lower two-thirds of Missouri, with the densest populations in the three counties surrounding Bull Creek—Christian, Stone and Taney counties. In bear management, there is a concept called the "carrying capacity," the amount of land needed to support a bear. But we now speak also of a "social carrying capacity," the number of interactions with bears that humans are willing to accept. As bear populations expand, so does their range, often into areas with dense human populations. Bears are opportunistic omnivores. Birdfeeders, trash, garbage, pet food and beehives all look like a smorgasbord to a bear wandering into what we like to think of as our personal property. Living in bear country requires controlling access to these delicacies. In theory, humans should be more trainable than bears. Having bears in the Bull Creek watershed is a good thing—a sign that the watershed, and its ecosystem, are largely intact and functioning properly.

Bear helping Itself to food in the bird feeder.
Photo credit: Steve Martin

Just as bears are important to watershed health on land, aquatic predators are important to the health of the stream. A good balance between predators and prey keeps the aquatic ecosystem working smoothly. In the case of Bull Creek, this ecological balance is maintained by a top predator in the aquatic realm—the smallmouth bass. Smallmouth are widely considered to be the premiere Ozark game fish. Pound for pound, they fight as good as anything out there. Unlike trout, which are not native to Missouri waters, smallmouth are long-time residents—our familiar, homegrown fish, inhabiting clear streams draining into the upper Mississippi and Ohio Rivers.

The scientific name of the smallmouth bass is *Microptera dolomieui.* The specific name, "*dolomieui*," is after a French scientist, Deodat de Dolomieu. He was a colorful character. When he was 18, he killed a man in a sword duel, landing him in prison for several months. In 1798, while serving as a scientist on Napoleon's military expedition to Egypt, his ship sank off the coast of Italy. He struggled to shore, only to be captured and imprisoned in Naples. Other scientists, even from countries at war with France, protested this rough treatment of a distinguished naturalist. Dolomieu's friends at the Academy of Sciences came to his rescue and secured his release. But the months spent in a dank prison ruined his health, and he died soon afterward.

It is fitting that smallmouth bass are named for Dolomieu—an exuberant, aggressive character. And it is also very fitting that some of the favorite haunts of smallmouth are streams like Bull Creek, lined with Dolomieu's other namesake, dolomite, which often rises in dark, towering bluffs.

In Bull Creek, bass find a familiar, comfortable home among limestone and dolomite boulders and bluffs. In large measure, the rocks and the fish together largely define the Ozark stream.

It is sometimes said that trees grow fish. This is because forested headwater streams serve as incubator areas and nurseries for a variety of fish and other creatures. Dead leaves, falling or washing into these streams, become food for a host of organisms. Leaves are made of lignin, a tough, fibrous material that is difficult for most animal stomachs to digest, but by the early spring, leaves clotting the stream—having been partially digested by bacteria and fungi—are black, slimy and soft.

Above: A Bull Creek smallmouth bass
Left: Tim Smith fishing in Bull Creek

This makes them even more digestible to the next higher organisms in the food chain, such as the aquatic nymphs of insects like stoneflies and caddisflics, which have mouthparts designed to shred and chew softened leaves. Fungi and bacteria imbedded in the decomposing leafy matrix provide these "shredders" with added protein. These aquatic insects, in turn, become food for carnivores such as frogs, turtles, snakes and of course, fish. Consequently, watersheds with good forest cover tend to produce healthy bass.

Below the point where Bull Creek emerges from the national forest, the landscape changes dramatically. Larger pastures line the stream and, in places, even the backyards of suburban homes. Sunlight pours into the stream zone and sun-loving algae grow profusely on rocks and the stream bottom. In many places, diatoms—miniscule, golden brown plants with silica in their cell walls—cover the rocks. They are as slick as snot, making it hazardous to navigate over flat rocks on the stream bottom. An abundance of diatoms and algae means more algae-eating fish like stoneroller minnows and suckers. Bottom line—forested sections of streams tend to produce more sunfish and

Caddisfly in its "case." Photo by: Jim Rathert, Missouri Department of Conservation

A.

B.

C.

D.

Macroinvertebrates:
A. stonefly larvae,
B. mayfly larvae.
C. water penny (beetle larvae)
D. hellgrammite (dobsonfly larvae)

bass; open, sunny sections tend to produce more stonerollers and suckers. There is nothing wrong with having these fish in the stream, of course, but when they're found in great abundance, it means the balance has shifted to waters that are overly enriched with nutrients.

Not surprisingly, hard-fighting smallmouth bass are a favorite quarry of Ozark anglers. But many people who fish for them claim that otters have decimated the fishing. Otters are now commonly seen in Bull Creek and other Ozark streams. But it wasn't that way several decades ago. Only about a hundred otters existed in the state in 1980, relegated to swamp remnants and sloughs in the Bootheel. The Missouri Department of Conservation (MDC) stocked otters from Louisiana into a stream in north-central Missouri in 1982, and over the next eleven years, introduced them into forty-three more streams across the state. The otters did exceptionally well, quickly dispersing from stocking sites, reproducing vigorously, and eating their way through fish and crayfish populations in creeks and ponds across the state.

Otters, of course, were here first. They are native to Ozark streams and were once found here in abundance. Because they are superbly adapted to life in streams, their numbers rebounded rapidly after re-stocking. Their successful return is another good sign that the Bull Creek watershed is healthy. However, large numbers of people are now competing with them for the same prey. Otters love fish, but also like crayfish. Some biologists think that otters have a competitive advantage over fish like bass in catching crayfish, which might also explain a reduction in the populations of fish.

Beavers, bears, otters and bass all respond to changes in the watershed. But in the past, most of the change was natural, brought on by slowly changing climactic conditions like the ice ages, which created a long period of cooling and drying. But there are now signs that people are having an impact on the rate of change—longer, hotter summers, for example, or bigger floods. Bull Creek, like other streams across the Ozarks, will be affected by climate change in ways that we can only speculate about at this point. But every organism, plant and animal alike, will be affected to some degree by the changes that are coming.

Relationships between organisms will also change. For example, warmer springs could induce flowers to bloom sooner, before critical pollinator insects are even around. Trees producing leaves sooner might prevent early spring ground plants from getting enough sunlight. Warm weather loving species like armadillos and water moccasins might move northward, displacing other animals. Cooler weather loving species like ferns and certain amphibians might disappear from the

Darters are found in cool, clear Ozark streams. Image courtesy of Missouri Department of Conservation.

watershed. Like its counterparts around the state, Bull Creek will serve as a living "test tube" for responses to the changes that we have brought upon it, and ourselves.

In the meantime, Bull Creek is still healthy enough to support a rich diversity of stream life. Darters, tiny fish wearing bright breeding colors in the spring, dart among the cobbles on the stream bottom. Channel catfish, goggle-eye, gar, and sunfish hover in deep blue waters. Wading birds like bitterns and herons spear small fish in the shallows, while turtles and water snakes bask on protruding logs. In the early spring, the river corridor is filled with the scent of witch hazel and a chorus of colorful songbirds. These wonderful sensations are still there, awaiting the interested and alert human visitor—these are what draw people to a thriving place like Bull Creek.

In a book called *Entering the Watershed*, the author described healthy watersheds as potential "refugia," or "healing centers." Bull Creek could very well be one of these—a healthy, relatively intact watershed—not pristine, of course, but containing one of the highest quality streams to be found. In fact, this is the reason the state of Missouri uses upper Bull Creek as a "biological reference stream," against which the health of other streams can be measured.

Polyphemus moth enjoying the shade on a summer day

But it is even more important as a storehouse of critical genetic information imbedded in an as-yet smoothly-functioning, mostly non-stressed natural system. As other watersheds around it become damaged and degraded, a healthy Bull Creek can serve as a refugia for organisms that are under stress elsewhere. It can also serve as a healing center, reinvigorating nearby degraded areas like a healthy organ sending out white blood cells to engulf pathogens in a diseased one.

That view from a high ridge or glade—an overlook of the healing center—gives one the satisfaction that this magical place remains mostly whole—largely intact. Hopefully, the view will not change significantly in the future—the stream will still flow, clear and bright, and animals and plants will go about their everyday lives, unencumbered by pollution or degradation or wholesale changes in the watershed brought about by excessive land-clearing or unwise development or overgrazing or careless recreational uses. But that future is in no way guaranteed. It will be up to the people who live and work and recreate in the watershed to make it happen, if they choose to do so.

Spring, Summer, Fall, and Winter on Bull Creek.

Projectile points and approximate periods of use

CHAPTER 3

FIRST PEOPLE

On the banks of Bull Creek, it is common to find little pieces of chert, a hard silica mineral often found in association with limestone. These smooth, angular, sharp-edged stones may not seem to deserve a second look. But when you pick up one of these stones and run your fingers over its sharp, serrated edges, you may suddenly realize that this artifact was produced by human hands. Then it is easy to imagine a Native American sitting on the riverbank, watching Bull Creek flowing by, working to perfect a stone tool for hunting or working hides or other chores.

Jack Ray of the Missouri State University Center for Archaeological Research has written a book about these ancient tools. Most of them, he says, are not arrowheads, but are more generically considered to be projectile points and knives, abbreviated as PPKs. They would have been attached to knife handles or spears or atlatl shafts long before the appearance of the bow and arrow about 1,400 years ago. One projectile point found near Bull Creek dates back 11,000 years.

The first people who walked and hunted the Bull Creek valley left few signs of their passing—no roads, or tracks, or permanent structures of any kind. The tribal names of the earliest people are unknown. Most of the names we now use were bestowed by the early Europeans explorers (e.g., Marquette and Joliet). At some point, members of the Osage Tribe moved into Missouri and at least temporarily inhabited Bull Creek country. Oral tradition suggests that the Osage nation formed in the lower Ohio River Valley. They migrated westward across the Mississippi River near the middle of the seventeenth century, perhaps pushed by the invasion of the Iroquois. Their main villages were approximately ninety miles northwest of Bull Creek, along the Osage River.

Near their villages, the Osage would plant their crops in the spring before leaving on hunting expeditions to the south and west. They would travel again in the fall, moving to temporary hunting camps in the White River valley. Schoolcraft saw their abandoned camps along Swan Creek, just east of Bull Creek:

> We passed successively three of their camps, now deserted, all very large, arranged with much order and neatness, and capable of quartering probably, 100 men each. Both the method of building and the

Osage Chief with Two Warriors 1861/1869 by George Catlin. Courtesy of the National Gallery of Art.

order of encampment observed by this singular nation of savages, are different from anything of the kind I have noticed among the various tribes of aboriginal Americans, through whose territories I have had occasion to travel. The form of the tent, or camp, may be compared to an inverted bird's nest, or hemisphere, with a small aperture left in the top for the escape of smoke; and a similar, but larger one, at one side for passing in and out. It is formed by cutting a number of slender, flexible green poles, of equal length, sharpened at each end, stuck in the ground like a bow, and crossing at right angles at the top, the points of entrance into the ground forming a circle. Small twigs are then woven in, mixed with the leaves of cane, moss and grass, until it is perfectly tight and warm. These tents are arranged in large circles, one circle within another, according to the number of men to be accommodated. In the center is a scaffolding for meat from which all are supplied every morning, under the inspection of a chief whose tent is conspicuously situated at the head of the encampment, and differs from all the rest, resembling a half-cylinder inverted. Their women and children generally accompany them on the hunting excursions which often occupy them three months.

Big Heart-Osage, 1930. [SHSMO Art Collection, B2175]

Schoolcraft began his 1818 trek in the late fall to avoid running into Osage hunting parties. The tribe had acquired a reputation for capturing settlers, taking their guns, traps and supplies, then releasing them, usually unharmed. The Osage knew the hunters would return later with a new set of supplies, thus setting up an involuntary commercial exchange network. Schoolcraft was told the Indians should have returned northward to their villages by the time he reached southwest Missouri that winter of 1818-19.

During his travels, Schoolcraft learned something of the Osage hunting practices, and how they compared with those of the settlers:

> The Indian considers the forest his own and is careful in using and preserving everything which it affords. He never kills more meat than he has occasion for. The white hunter destroys all before him, and cannot resist the opportunity of killing game, although he neither wants the meat, nor can carry the skins. I was particularly struck with an instance of this wanton practice, which lately occurred on White River. A hunter returning from the woods heavy laden with the flesh and skins of five bears, unexpectedly arrived in the midst of a drove of buffalo, and wantonly shot down three, having no other object than the sport of killing them.

The Osage were very athletic and robust. Several sources describe them as tall and powerful, with many exceeding six feet in height. The tribe was the dominant power throughout the Ozarks until pushed out by westwardly migrating Euro-Americans. In 1808, William Clark, who became an agent for Indian Affairs after his 1804 expedition with Meriwether Lewis, signed a treaty with the Osage, forcing them to move westward. But the Osage didn't interpret the treaty to mean they were no longer allowed to return and hunt on the lands they had "given up."

At about the same time, other native tribes, such as the Delaware, Kickapoo, Cherokee, and Shawnee, were being forced westward by treaties as settlers began filling the Missouri Territory. By 1812, about five-hundred Kickapoo had settled at a temporary village in what is now southern Springfield. In 1818, members of the Delaware tribe gathered around the trading post of William Gillis, located on the James River near the present-day city of Nixa. Piankashaw and Peoria settlements formed where Bull Creek joins the White River. A few years later, in the late 1830s, the Cherokee passed through the newly incorporated town of Springfield, thirty miles north of Bull Creek, on their forced march, the infamous "Trail of Tears," before heading southwest into Oklahoma Territory.

The Osage remained the dominant group of Native Americans in the Ozarks until the treaty of 1825, when they ceded their remaining lands in Missouri and Arkansas and were forcibly moved to Kansas. However, they still returned to hunt in the Ozarks from time to time until the end of the nineteenth century, creating conflicts with other tribes as well as with settlers. By the 1830s, both Indians and settlers complained about the lack of game for food and the reduced numbers of bear and beaver that had formerly been their basis of commerce.

In his journal, Schoolcraft painted an idyllic picture of Native Americans living in perfect harmony with their environment. But with better historical information, this image would not hold up. As they were forced west of the Mississippi by the government, Native Americans encountered other tribes on the lands that they had been promised. The competition for land and game was fierce, resulting in intertribal conflicts, stolen horses, and captured slaves. The appetite for trade goods such as metal tools and guns led to over-harvesting of animals beyond the tribe's immediate needs. The carcasses of many animals killed for their hides were left rotting in the sun, much as the settlers were doing.

Relationships between the crowded tribes and invading settlers were often complex. The Delawares settled along the James River, "where they fell prey to the usual horse-thieving whites." Traders carried out a booming business in liquor sales and charged exorbitant prices for corn, even as the Native Americans were starving. Historian Elmo Ingenthron describes their plight:

"By the time the emigrant Indians reached the Ozarks they were living much the same as the early settlers. They wore hats, coats, pants, and other clothing similar to the clothing of the whites. They usually decorated their adopted clothing with shiny objects, beads, and bright colored trimmings. They had learned to use guns, metal tools and copper kettles. They grew corn, pumpkins, beans, and squash. They built log cabins like their white neighbors. In fact, some of their own tribesmen were white men who had married Indian squaws and arrived in the Ozarks with the full-bloods and their mixed-blood children."

We now know that Native Americans manipulated their environments in significant ways. The open-view landscapes seen and recorded by Schoolcraft, with grassy patches among scattered tall trees, were created and maintained by Native Americans. They routinely used fire to open woodlands and stimulate sun-loving grasses, thus providing good grazing for the animals they hunted. They left other signs of their passing as well. Many of the highways we use today, such as I-44, largely follow the same alignment as the trails of Native Americans—usually the most direct and efficient cross-country routes.

When Schoolcraft encountered the Osage camp on upper Swan Creek in 1819, neither he nor the departed Osage could have foreseen the changes the next 200 years would bring. The first century after Schoolcraft's visit would bring automobiles, electricity, new industries, labor-saving appliances, and communication networks. The next would bring jets and daily international travel and computers and the instant transmission of information via the Internet.

The latter-day settlers on Bull Creek, with their new technologies and knowledge, often possessed a heightened sense of power, privilege and even prejudice. They developed the capacity to alter Bull Creek in ways from which it might never recover. But they also possessed a newfound ability to understand its complex nature and identity—and to help it remain vibrant and healthy for the foreseeable future.

Henry Rowe Schoolcraft

CHAPTER 4

THE SETTLERS

Like the Native Americans, most of the early European visitors and settlers in the Bull Creek valley left very few, if any, traces. Hunters and trappers roamed the valley before 1800, and a few of them left their names for us to ponder—such as the beaver trapper, Findlay, for which nearby Finley Creek is named. (Why it's spelled "Finley," rather than "Findlay," we'll probably never know.) The early settlers—the first people who came with the intent to stay—would be guided and influenced by the topography of Bull Creek. There were trails in the region but no real roads, so at first most settlers arrived on flatboats traveling upstream on the White River. By the 1820s, there was a thin chain of cabins and small settlements along the White River north of Arkansas, extending upstream to near the mouth of Bull Creek.

Cabin sites were dictated largely by the lay of the land, with river valleys and access to good water hallmarks of desirable property. The settlers, in turn, began to mold the topography to fit their needs, with land clearing and plowing for crops and tree cutting to create pastures for cattle. The effects of those activities on the watershed, though in some cases locally severe, were negligible in a watershed still mostly comprised of untouched wilderness.

Some of the earliest settlers along the upper White River, in the vicinity of Bull Creek, were visited and described by Schoolcraft. With his friend Levi Pettibone, Schoolcraft left Potosi in November 1818. The pair traveled south from Potosi toward the southwest corner of the Missouri Territory to check out the reported lead deposits on the James River. On the way, they stopped at the home of the Wells family, who had just built a cabin on the White River. Schoolcraft described their new log-house in condescending terms:

> Its interior would disappoint any person who has never had an opportunity of witnessing the abode of man beyond the pale of the civilized world. Nothing could be more removed from the ideas we have attached to domestic comfort, neatness, or conveniency, without allusion to cleanliness, order, and the concomitant train of household attributes, which make up the sum of human felicity in refined society.

Of the Wells children, he remarked:

> The boys were clothed in a particular kind of garment made of deerskin, which served the double purpose of shirt and jacket. The girls had buck-skin frocks, which it was evident, by the careless manner in which they were clothed, were intended to combine the utility both of linen and calico, and all were abundantly greasy and dirty.

After leaving the Wells' cabin, Schoolcraft and Pettibone ascended the White River to the mouth of Beaver Creek, a few miles east of Bull Creek. Here, they met the Holt and Fisher families, said to be the most western settlers on the White River. They had arrived only four months earlier and had not even completed their cabins. Though newly arrived, they already knew the names of Bull Creek, Swan Creek, and the Findlay Branch of the James River, indicating these names had been in common use by hunters or trappers before they got there.

Schoolcraft published the account of his Ozark adventures in London in 1821. While he described many of the settlers and their cabins in extremely unflattering terms, he painted a rosy picture of some parts of the region and the potential for commerce and settlement. The valley of the James River, he said, had "extensive bodies of the choicest land" and "rich, black alluvial soil, apparently deep, and calculated for corn, flax, and hemp." Boats of "eight tons burthen" should be able to ascend the James at least to the mouth of the Finley, "establishing a communication by which the peltries, the lead, and the agricultural products of the country, could be easily, cheaply, and at all seasons taken to market, and merchandise brought up in return."

Elizabeth McCoy gravestone in the Bull Mills cemetery

Armchair explorers and would-be settlers no doubt read Schoolcraft's travel account with interest. But a much larger inducement for settlement had already occurred a few years earlier, in 1816, when Congress passed legislation opening land west of the Mississippi for settlement. A slow but relentless wave of families seeking land and opportunities began moving across the Missouri Territory. Most of the settlers of the interior Ozarks, including the Bull Creek country, came from mountainous states such as Kentucky and Tennessee, drawn to hills and valleys reminiscent of their homelands.

Early settlers on Bull Creek journeyed into the unknown with the barest of tools for survival. Settlement sites were chosen carefully, with dependable water supplies of primary concern. Especially sought out were perennial springs issuing onto river benches above low-lying areas, which were feared for their "malarias" and "miasmas." Coming from the hills, the pioneers were predominately woodsmen wielding a plow rather than flat land farmers. But it was important for all settlers to have enough level, fertile ground to plant a few crops, usually in floodplains or on benches flanking a stream. Settlers often gauged the quality of soil for farming by the size of the trees growing on a site. The bigger the trees, they assumed, the better the soil. Tall, thick hardwoods near the stream signaled to them areas that, with clearing, would yield healthy crops of corn, wheat, or flax. So, for an ideal spot to put down their roots, the best sites contained at least a portion of the valley floor, the so-called "bottomlands." Schoolcraft described one such site chosen by settlers along the White River:

> They have fixed the site of their habitations on the east banks of the river, on the verge of a very large and rich tract of bottom land, occupying a bend in the river. It is covered by a heavy forest of oak, ash, maple, walnut, mulberry and sycamore, the latter skirting the immediate banks of the river, with a vigorous growth of cane below. The opposite bank of the river is a perpendicular bluff of lime-stone rock, rising at the water's edge to a height of 300 feet, where it terminates in very rugged peaks, capped by a stinted growth of cedars and oaks, and forming a most striking contrast with the level, rich, and heavy wooded plain below, over which it casts its broad shadow by half-past three in the afternoon, which must render it a cool and delightful residence in summer."

The settler's first task was to fell trees to make a cabin. Trees near the cabin site were killed by girdling them. Some standing dead trees could be taken down later for fences and firewood. Others could be cleared sooner to make way for plowing a garden patch and planting a few crops, especially corn. Wheat was grown for white bread. Tobacco, flax, hemp, and cotton were grown for home use. A cow would be added for milk and hogs for meat. Turkey and deer were plentiful in the surrounding woods, and foraging for wild greens, berries, fruits, and nuts helped round out the diet.

1830's mill stone from Bull Mills

In the early years, homesteaders simply gathered or raised what they needed or traded some of their produce with neighbors. Although the homesteaders might be millers or chair makers by trade, they would still raise much of their own food.

At first, when the numbers of settlers were very small in comparison to the region's size, the impacts of their activities were minimal. But over time, as more and more people moved into the watershed, some deleterious effects began to be felt. The felling of trees, especially in floodplains, opened these areas to the erosive force of floods. Grazing on steep slopes or on thin, rocky soil decreased vegetative cover, allowing erosion to remove much of the remaining soil. Cornfields required plowing, also exposing soil to the effects of erosion, especially on sloping ground. Of course, the earliest settlers, who could only see the smoke from a distant neighbor's fire, gave no thought to how their simple working on the land could damage the vast forests surrounding them, or cloud the clear stream at their doorsteps.

Geographer Carl O. Sauer wrote extensively about farming in the Ozarks. In the early 1900s, he noted that corn was the "traditional crop of the hill farmers." The rugged White River Hills, at the lower end of Bull Creek, had some of the roughest topography and poorest soils in the state; but corn would grow there, and farmers relied heavily on it. It was well suited to subsistence farming and could be used directly as food or ground into meal for cornbread. Any excess could be fed to livestock or made into whiskey. Shucks could be made into mattress stuffing and brooms and cobs could be used for fuel.

The need for grinding corn into meal or wheat into flour provided an early impetus for the establishment of grist and flour mills. James Kimberling arrived on Bull Creek in 1832 and, with his two brothers, built a water-powered grist mill, one of the earliest in the region. The mill was about five miles above the mouth of Wood's Fork Creek in present-day Christian County. This structure would eventually give its name to a tiny community called Bull Mills (now on the property of Bob and Barb Kipfer).

Mill dams on streams were usually made of rock and masonry, and high enough to back water a few hundred yards upstream. The miller only needed enough "head" of pressure, or vertical distance between the top of the dam and the stream surface below the dam, to turn a wheel or turbine. With such a small mill pond, and with most of the water flowing over the dam, the ecological effects on the river were probably minor. As a positive benefit, the mill pond provided a great summertime swimming hole and picnic site.

Though most settlers grew corn, it was worth only 30 cents a bushel at the turn of the twentieth century. With the lack of good roads, hauling excess corn for sale elsewhere was seldom profitable. Some usually ended up going to the hogs along with the slop. On the other hand, making moonshine from corn mash could provide a subsistence farmer with some cash to purchase necessities. With a still and a small investment of time and labor, a man could make whiskey worth over $2 a gallon. Whiskey thus became a more portable and profitable product than corn sold in bulk.

Ozark postcards often show hillbillies making moonshine in some remote valley, relaxing by their still, smoking their pipes. There was some truth to this. There was also a lively tradition of the moonshiners' attempts to evade the "revenuer." Taxes on alcohol date back to George Washington's time and were never popular. An 1862 law required a tax of $2.00 per gallon, which could wipe out the meager profits of small-scale distillers. For this reason, clandestine stills became popular in the Ozark hills. Since moonshiners went out of their way to evade detection, there is no telling how many stills operated in the hills around Bull Creek.

During the Prohibition years, the stakes were even higher for illicit whiskey-making. But laws were enforced with varying vigor, depending on the location. Neighbors generally turned a blind eye to those working local stills. Avoiding taxes remained illegal even after Prohibition ended, but that didn't stop many distillers from making a little "shine." Today, in remote sections of the Ozarks, a person may sometimes come across the remains of a still, usually far away from houses in a narrow ravine. Often, water for making moonshine came from a nearby spring, which may still contain lengths of rusty pipe. The environmental impacts were in most cases minimal, other than the remaining "trash," which to some people could be of historical interest.

As the population along Bull Creek increased, farming for cash became an occupation as well as a way of life. Farmers obviously needed enough good soil to grow marketable amounts of crops. Some ridgetops and narrow bottomland areas had deeper, tillable soil, but steep slopes and thin, rocky soils in other areas presented formidable constraints on productive farming. Sauer observed in about 1910 that the "greater part of the Ozark Highlands consists of unimproved, so-called wild land, covered with forest or brush." This was especially true in the southernmost part of the Missouri Ozarks, including Bull Creek.

Land in the Bull Creek watershed had a range of suitability for farming. The northern part, with grasslands and level ground, was much more productive for crops or livestock. In southern parts, particularly the White River hills, the soil was thin and rocky. This was reflected in the area of "improved" farmland tabulated by Sauer at the south end of the Bull Creek watershed in Taney County. In 1910, 16% of the land in Taney County was in improved farms, while in Greene County, to the north in grasslands, 71% of the land was in the improved category.

The percentage of land in farms was also much smaller to the south. Taney County had less than 40% of its land in farms, while Christian County, containing the northern two-thirds of the Bull Creek watershed, had over 60%. The average size of farms was also larger in Taney County, since it took a greater amount of hilly land to produce the same level of marketable crops. The average Taney

County farm was over 150 acres, while in its northern neighbor, Christian County, the average size was less than 100 acres.

The smaller number of farmed acres in Taney County reflected the fact that cultivation of hilly land required more labor and was less productive per acre than flat prairie or bottomland. But it was also difficult to get products to market from remote sections like those at the southern end of Bull Creek. As a result of these factors, Taney County in 1910 had one of the lowest land values of any county in Missouri—less than $10 per acre. In contrast, the value of land in Greene and Christian Counties was $25 to $50 an acre.

Canebrake: heavy growth of southern giant cane

While general farming would become predominant in the flat upland prairies in the latter parts of the nineteenth century, the hill country of Bull Creek remained largely at the subsistence level, with little in the way of cash crops. Lands mostly wooded were more conducive to livestock raising than to farming crops. Free range hogs and cattle became the norm. With free range, the land is not fenced, so it is available for anyone's stock to use. Sauer noted that by 1910, open-range livestock grazing had caused much of the once productive "bluestem pasture" to become of extremely poor quality, "especially for cattle."

Early farming techniques probably took a toll on the water quality of Bull Creek. Clearing bottomlands of trees to make room for pastures created long term erosion problems. Cattle trampled bottomland vegetation and overgrazed grasses, trees and bushes in floodplains and on gravel bars. Hogs rooted along the stream and wallowed in streamside mudholes. The dense streamside canebrakes described by Schoolcraft, eagerly sought for fodder by cattle and hogs, would soon disappear. The increase in bare ground led to much greater erosion in the riparian zone, the vegetated zone along the stream.

Without good vegetative cover, streambanks eroded laterally, scouring sediment from the outside of bends and depositing it on the insides, causing the channel to migrate across its floodplain with every major storm event. The increased erosion of banks and adjacent land released gravel that had accumulated in streambanks and floodplain sediments over thousands of years. Huge pulses of gravel now moved downstream with every major flood event. Recent studies in the Ozarks by the United States Geological Survey indicate that riparian zone damage from early agriculture may have led to more gravel being deposited in streams than early forestry practices did.

Feral hogs in their wallow

By rooting and wallowing near streams, hogs also caused significant erosion and water quality damage. Hogs were once prominent in the Ozarks. The settlers Schoolcraft encountered in 1818-1819 invariably raised a few hogs. Omnivorous and self-sustaining, hogs fed not only on acorns in the forest, but also on insects, worms, eggs of ground nesting birds and occasionally, a newborn mammal.

Fences were built around crops to keep the roaming hogs out, not in. In season, acorns and other foods on the forest floor was all the forage the hogs required. The expression that "even a blind hog finds an acorn once in a while" comes directly from their love of this food source, so abundant in the Ozark forests.

Widely ranging hogs and cattle began to create problems, but fencing laws were slow to reach Bull Creek. On one occasion, hogs terrorized settlers in Forsyth, just south of Bull Creek. They raided gardens so often that vigilance committees began patrolling the area. Historian Elmo Ingenthron described an episode in 1896 where hogs mutilated a corpse retrieved from a hangman's noose. The body was laid in front of the courthouse to serve as an example to others. A newspaper reported on the subsequent city election where the citizens elected "one hog and four anti-hog councilmen. Forsyth is very progressive."

Fencing laws finally put an end to free-ranging livestock. But other porcine problems were already afoot by then. Beginning in the 1890s, pure Eurasian wild boars were introduced into several areas of the U.S. to provide a big-game species for wealthy hunters. These hogs were mostly confined in ranches, but there were inevitable escapees. Still, for many years, populations of feral hogs were small. In the 1960s, hog populations were believed to exist in only three small areas in southeastern Missouri. An estimate of the population at that time was less than a thousand animals.

In 1969, the Missouri state legislature finally enacted a closed-range law outlawing unconfined livestock. As a result, the numbers of feral swine in the state were reduced. But the rise of hog hunting for sport in the 1980s led to the intentional and illegal release of hogs onto public lands. As hog hunting became more popular, raising wild boar as a form of alternative agriculture for hunting on licensed shooting areas became a more lucrative business, and escapes and even deliberate releases were common. The populations of feral hogs began to increase.

Adding to the problem, the domestic hog market crashed in the early 1990s after large slaughterhouses began bypassing open markets and buying pigs through private contracts with growers. With fewer buyers in the market, prices sagged. Some farmers simply released their domestic pigs into the wild. Since feral swine are very prolific and adaptable, it didn't take long for their numbers to increase across a large portion of southern Missouri.

By the end of the 1990s, multiple Missouri agencies were receiving hog damage complaints. According to a recent USDA report, feral hogs cause $2.5 billion in damage every year, half of which is agricultural damage. In addition to eating everything that gets near their mouth, they root in the soil of farms, yards, and golf courses, earning them the title of "the Rototillers of Nature." Damage to natural areas includes loss of native plants, competition with other animals for food, and soil erosion from wallows and trails. Hogs need to wallow in water to cool off, which can lead to their waste contaminating streams. They are a vector for human as well as swine diseases. Transmission of disease could devastate hog farms in the U.S. as it has in Europe and Asia.

Feral hogs are breeding machines. A sow can deliver two litters (average of six piglets per litter) in fifteen months, and females can start reproducing at five to six months of age. Simply put, feral hog elimination is a herculean task. Today, federal and state agencies are working vigorously to eliminate feral hogs by trapping and even shooting them from helicopters. But there are still fenced game farms in Missouri for hog hunting, with the inevitable escapees. In the wild, hogs usually move together in a small herd, called a sounder. Hunting is mostly ineffective, as shooting one hog simply drives the rest of the sounder away from baited traps.

In addition to hogs and cattle, fish were once raised in the Bull Creek watershed. Most fish farms raised trout in springs or spring-fed ponds for stocking in streams or private lakes. At least one grower raised goldfish for pets to put in glass bowls, a fad started by pet departments at variety stores such as S. S. Kresge and Woolworths. Fish hatcheries also supplied goldfish and minnows for bait. A hatchery on the West Fork of Bull Creek was operated by the Glenstone Bait Company. In 1948, the owner harvested 700,000 bait minnows, along with thousands of crayfish.

Trapped! Feral hogs are baited and caught in a trap.

All the hatcheries in the Bull Creek watershed are now defunct, at least as far as raising fish. Most of the ponds are dry, and old dams and flumes are cracked and crumbling. It is difficult to know what environmental impacts these facilities might have had when they were in operation. Dams reduced flows downstream, which would have affected aquatic habitats and wildlife. As shallow ponds in sunny areas heated up, they would release warmer water downstream. This could have reduced populations of cold-water adapted species, like darters and sculpins, living below the pond. Of course, when the fish farms near Bull Creek were in business, none of this was considered by the owners or anyone else. Recently, however, rules have been put in place to minimize the environmental impacts of large fish farms.

Some of the old fish farms now form scenic amenities for residential properties. Nice homes have been built near ponds, which have been stocked with trout or bass. At other old fish farms, where ponds are now empty and dams are gone, trees or weeds are growing where fish once swam. Many other fish farms around the state have closed over the last few decades. Owners and former owners claim that to stay in business, high volumes of fish or bait must be sold. Apparently, the Bull Creek farms were never large enough to compete successfully over the long term.

Bull Creek farmers experimented with other agricultural enterprises over the years. At the turn of the twentieth century, just as the timber trade was dying, extension agents promoted tomato growing. Soils in the Bull Creek watershed were compatible with tomato production, and the whole family could contribute to the labor of planting, tending, and harvesting. Tomato canneries sprang up in many communities, many of which were family-based, seasonal operations that also provided jobs for women and children. Drought and blights in the 1930s ended the tomato boom and before much longer, the canneries closed.

Other agricultural schemes also raised expectations but didn't pan out in the long term. For a few years, growing strawberries became popular. Apple and peach orchards were also tried. In 1924, agricultural promoters convinced farmers in the Bull Creek area that there was big money to be made raising onions. Many farmers jumped at the chance. Bumper crops of onions were produced, filling all the available barns and garages. When the growers discovered that there was no market for this produce, they raised a stink, like the onions rotting in their barns.

All these excursions into growing alternative crops had their drawbacks. Without annual applications of manure or commercial fertilizers, thin, rocky soils soon became depleted of nutrients. Crop yields gradually diminished after the first few years. And as with fish farms and cattle farms, smaller operations always had a tough time competing with bigger operations elsewhere, especially those which had ready access to larger markets and better transportation facilities.

Cattle ranch in the Bull Creek valley

In the 1970s, a new subsistence crop appeared—marijuana. As far as we know, pot never became a major commercial product along Bull Creek. Most of the growing was no doubt small-scale for personal consumption, although it is likely that some was traded for other products, just like corn and moonshine had been items of barter nearly two-hundred years ago. Many people ended up with a criminal record for possessing pot in amounts that are now perfectly legal.

It is difficult to say whether marijuana patches exerted any deleterious effects on Bull Creek ecosystems. Most patches were small—and therefore easier to hide from prying eyes. But in the 1970s, during the "War on Drugs," some patches were sprayed with herbicides upon discovery by law enforcement agencies. The resulting product, if smoked, could certainly have had ill effects on the user.

These days, the most prominent agricultural enterprises along Bull Creek are cattle operations. Where soils are fertile enough to support them, pasture lands spread across the landscape, most of them on flatter ridgetops or in the bottoms of valleys. The days of free-range cattle are over, of course, and most farmers now use better practices than their forebears. However, in many places, cattle are still allowed unlimited access to streams and spring branches. The result is often overgrazing and trampling of streamside vegetation and high loads of manure in or near the stream, ready to be flushed downstream with runoff.

We now know that grazing, if done carefully, can benefit soil health and be protective of water quality and streams. One method being used is managed or rotational grazing, where pastures are divided into cells, or paddocks. Animals graze in one cell while the others are resting, allowing time for the growth of new grass in resting cells. As cows rotate through cells, manure is spread more evenly across pastures rather than concentrated in stream zones. When cattle enter a fresh cell, they get to eat the most tasty, nutritious part of the new grass blades. Weight gains are better, and the farmer's profits fatten. Still, because these systems require more intensive management, owners may be reluctant to invest time and money in them.

Things have certainly changed from the days when a Bull Creek resident was utterly dependent on harvesting wild game and coaxing a few corn plants from newly broken soil. While there are still many farms in the Bull Creek watershed, most farmers these days have alternative means of support. Many have regular jobs in the surrounding cities or towns, and "run a few head" of cattle on the side. Because farming may no longer be a primary source of income, less attention may be paid to farm management, including a focus on adopting better practices. Encouraging and promoting these practices is a process of education and persuasion.

Today's Natural Resources Conservation Service, an agency of the U.S. Department of Agriculture, has many recommendations on how to farm without creating erosion problems or polluting streams. Whether farming with crops like corn or providing pastures for grazing, one of the most important practices is to leave a wide, vegetated buffer strip along any stream or drainageway. A buffer strip will absorb and filter any animal waste or fertilizer that might be in runoff. It is also important to maintain healthy stands of forage for cattle, like warm season grasses, which will prevent erosion and runoff of soil and animal wastes. Applying fertilizers can be done in ways to prevent runoff into streams, which can stimulate the growth of nuisance algae. With good farming practices, that bucolic scene of productive green pastures surrounded by healthy forests can remain rooted in reality, as well as so pleasingly in our minds.

Galena, Missouri Mines State Historic Site

CHAPTER 5

GETTING THE LEAD OUT

On Bull Creek headwater streams, a hiker will sometimes come across small craters in the soil or rough humps of rocks and earth. These are exploratory pits from the old lead mining days. Examining rocks around these diggings will usually reveal silvery fragments of embedded galena, or lead ore. There is not much lead here; no doubt a disappointing result for the hopeful miners who long ago scratched into these hillsides. It is difficult to imagine how anyone ever made a decent living, much less struck it rich, mining lead from these rugged hills. Such futile efforts might represent another example of Samuel Johnson's "triumph of hope over experience."

Mining is one of the oldest commercial activities in the Ozarks, as elsewhere, and was at the heart of early European exploration. In the United States, it was one of the principal drivers for westward expansion. Early mining for lead was a byproduct of the mostly unsuccessful search for gold and silver, which many people thought would be found in conjunction with deposits of lead. But lead itself had important uses, especially for the military. The French operated a lead mine in Missouri along the Meramec River in 1700. By the late 1700s, lead mines were active in the region northwest of Ste. Genevieve.

Schoolcraft was drawn westward from New York by reports of lead deposits in Missouri Territory. Arriving at Potosi, Missouri in late 1818, he stayed at the mansion of Moses Austin, who had arrived in 1798 to mine lead at Mine à Breton (later called Potosi). In his book *A View of the Lead Mines,* Schoolcraft described the crude mines and furnaces around Potosi. What was needed, he said, was an "infusion of capital" and "scientific knowledge and practical skills in boring, blasting, draining and ventilating." Schoolcraft's book whetted the appetites of the next generation of miners back east, some of whom went westward to try their luck in the hills of southern Missouri.

While in Potosi, Schoolcraft also learned about lead deposits on the James River in the southwest part of the territory. He and a partner set off from Potosi on a three-month expedition, a journey described in his book, a *Journal of a Tour into the Interior of Missouri and Arkansaw.* Eventually, he found his way to the lead deposits on the banks of Pearson Creek, a James River tributary, a few miles southeast of the present-day city of Springfield. He spent a few days there, making a crude smelter and looking for galena, which he spotted from its silvery reflection in streambanks and on the stream bottom. His return trip took him through our valley of Bull Creek, where he became lost for a day.

Southwest Missouri would become a big producer of lead, but not the area around Bull Creek. The "Tri-State" lead mining area west of Bull Creek included portions of extreme southwestern Missouri, southeastern Kansas and northeastern Oklahoma. The Tri-State Mining District would yield a lot of lead after its discovery there in the 1840s. Production here, added to that of the "Old Lead Belt" near Potosi and the "New Lead Belt" in the eastern Ozarks, allowed Missouri to produce over 90% of the lead in the U. S. by 1990. Southeast Missouri, site of the New Lead Belt, became recognized as a geologic wonder, containing the largest concentration of galena in the world.

Lead and zinc ores tend to occur near or even at the surface in the Ozarks. Deposits near Bull Creek were concentrated in ore but small in extent, limiting the feasibility of commercial mining. In the early days, every phase of extracting and purifying the ore required tiresome manual labor. Digging out the ore was backbreaking work, as described by Schoolcraft in Potosi:

> The method of raising the ores and the processes pursued in separating the metal ore, upon the whole, are extremely simple. A pickaxe and shovel are the only tools in use for removing the earth, and the drill, rammer and priming rod are added when it is necessary to blast. Having determined the spot for digging, the process commences by measuring off a square of about 8 feet, and throwing out the earth, spar and gravel, until the miner sinks beneath the depth (where) he can throw the earth. A practiced hand will pitch his earth clear out of the pit from a depth of ten, twelve and even fifteen feet. At this depth, a common windlass and bucket is placed over the center of the pit and the digging continued by drawing up the earth, spar, and ores, if any are found, in the manner pursued in sinking a well.

Once the ore was above ground, extracting the lead required a long, tedious smelting process. Though crude, the early furnaces were large enough to smelt over two tons of ore in a single charge. They were built on a slope, with a stone wall on the downhill side and lateral walls on either side, forming a sloping "U." Logs were placed in the furnace, the galena ore was piled on the logs, then more logs were laid over the ore. After stacking fuel and ore, the logs were set afire. The first wave of intense heat drove the sulfur off the ore. As the fire got hotter, the galena began to melt into liquid

A crude lead furnace. Diorama, Missouri Mines State Historic Site

lead. Roasting took many hours before the molten lead ran downhill into a pool at the front of the furnace. The leftover part of the ore, called slag, still contained about half the lead, but that was about the best that this crude smelting process could produce.

The lead smelting furnaces went through a lot of wood. Cutting, hauling and stacking the logs was hot and dirty work, and over time wood had to be hauled from farther away, adding to labor costs. Then there

The Armstrong "diggings"

was the loading and hauling of the lead, a very dense, heavy metal. Diggings in the Bull Creek valleys were mostly far removed from the railhead in Chadwick over rugged hills, making transportation difficult. Adding to the uncertainty of profitability, the price of lead fluctuated widely over time.

A man named Porter started the first mining venture in the headwaters of Bull Creek in 1853. By 1856, he had become a partner in the Bray mines on the East Fork of Bull Creek, just west of present-day Oldfield. An improved smelter and multiple shafts at this site fueled excitement about the future of lead mining here. The *Springfield Tribune* published a long piece about mining in December 1857, including this enthusiastic report on the potential for lead in Taney County:

This county is considered to be one of the richest in the state in mineral resources. Lead ore is found there in abundant quantities. In every instance where the miner has sunk his shaft, he has found the mineral. Mr. Bray is working his mine successfully. He is raising a large amount of mineral and with but very few hands. Capital is all that is now wanted to develop the richest lead diggings in the United States. Taney County will be a paradise to the miner.

Bull Creek Mining Company operated mines around the headwaters of the West Fork of Bull Creek, remnants of which are visible today. Any financial success this company might've had ended about 1890, with only a brief reprieve brought on by inflated lead prices during World War I. Downstream from the Bray mines, there are multiple pits five to ten feet deep. These diggings and others were not financially viable, given the low volume of ore and poor access to transportation, and so were deserted.

The Armstrong Diggings were about one and one-half miles north of Bull Mills, on a steep slope above an intermittent stream. An 1894 government report said of these Barber Creek Mines, "a milling plant has been erected and much money had been spent on the strength of mere future possibilities." Additional sites upstream were deemed "worthy of attention." At the site today, there is no evidence of the road or equipment that must have been needed to bring the ore up the long slope to the ridgetop.

Beginning in about 1920, churn drilling was commonly used for mineral prospecting. This resulted in the discovery of deeper deposits of zinc sulfide above Brown Spring, in the headwaters of the West Fork of Bull Creek. Zinc was in great demand to support the war effort during WWI. In the Bull Creek watershed, the Mary Arnold Mining Company built a gravity-concentration zinc mill with a daily capacity of two-hundred tons. The operations ended when reserves were depleted in October 1942.

Many promotional schemes were built on the promise of highly productive mines. This potential for wealth was dangled out in the press. Quick money was to be made by selling "promising" land to distant speculators, who frequently would never set foot in Missouri. Newspaper ads, such as this 1891 piece from the *Taney County Times,* were glowing:

Boom! Boom! Boom! We have mineral lands of all kinds, lead, zinc and silver, also onyx and all kinds of building stone for sale or lease. No property will be offered by this company until the character of its mineral deposits has been fully established.

Stump Spring

Much of the land near Bull Creek was owned by companies speculating for zinc and other minerals. Disappointingly for one prospecting party, a test boring along the Mail Trace Road produced only water. The company, not wanting to waste the effort, put a steel pipe and pump in the borehole to provide water for the community. Sometime later, someone with a grudge packed the pipe with dynamite and blew it wide open, leaving a large split. The remaining split pipe still oozes water and has forever since been called "Stump Spring."

The legacy of lead mining in Bull Creek, as in other lead producing areas, has been the contamination of soil from lead mining residues. In some cases, tailings from diggings and mines have washed into streams, causing lead to accumulate in stream sediments. After years of mining in the Bull Creek watershed, the levels of lead in stream sediments were most likely elevated. Riparian trees like sycamores and aquatic plants like water-willow can pull lead from sediment. In the fall, their lead-laden leaves fell into the stream and were eaten by primary consumers like caddisfly larvae and crayfish. Lead can build up in the bodies of these invertebrates, or be "biomagnified," and when fish or birds eat the invertebrates, lead can build up in *their* bodies.

Recent studies have shown that there has been very little biomagnification of lead in the tissues of Bull Creek organisms. This shouldn't be too surprising since lead mining in the headwaters of Bull Creek didn't last all that long and ended over one-hundred years ago. Most of the mining scars have healed, and the levels of lead in plants and animals have decreased over the years since mining ceased. Any lead that accumulated in the tissues of aquatic creatures in the past has most likely dissipated. Other parts of the state, such as the Big River in the eastern Ozarks, are not so fortunate, as the levels of lead in sediment there remains high.

Few reminders of the mining days remain in the Bull Creek watershed. Most residents today probably have no idea that the area has a long history of mining—that prospectors tried their luck, digging here and there, hoping to strike it rich. Diggings in the area were generally small, and the scars on the land have mostly healed. But the final chapter of the lead story may not have been written. Without more study, it is difficult to know if lead mining residues remain in the area's soils and sediments, and whether these pollutants might still be affecting some of the animals and plants that live there. Answers to these questions remain buried for now, awaiting a new wave of discovery.

CHAPTER 6

FORESTS

Forests, the most prominent natural feature in the Bull Creek watershed, have long provided the land a resilient layer of protection. Leaves absorb the impact of raindrops, as opposed to hard surfaces like concrete or rooftops, where exploding raindrops immediately dislodge and mobilize pollutants. Sediment is removed as runoff is slowed and filtered by a thick blanket of fallen leaves—a lofty, fibrous blanket that also holds tons of water, like a massive sponge. As a result, forest streams, as opposed to urban streams, rise much more slowly after the start of rainfall and flow longer after rainfall stops, as the leaf mass slowly fills and then drains. Streams in the Ozarks with extensive forest cover, like Bull Creek, are amazingly clear most of the time.

Varying landscape conditions in the watershed produce different forest types. Dry upland areas feature primarily oaks and hickories. Interspersed with them are dogwood and wild plum trees, providing splotches of brilliant white in the spring. Trees protect the stream more directly when they grow in the riparian zone, the ribbon of bottomland vegetation abutting the streambanks. A diversity of plants can be found in the riparian zone along Bull Creek—several species of trees, of course, but also river cane, woody shrubs and sedges, rushes, and grasses. Common riparian trees include sturdy green ash trees, along with hardy maples, gloriously red in the fall, and sycamores, with chalky white bark so starkly beautiful in the winter.

Trees were important to the settlement of Bull Creek. Homes, barns, furniture, fences, boats, and farm implements were all once products of the axe and saw. Cord wood provided heat for homes until coal or oil heating arrived. Most sawmills were steam-powered, using cordwood for boilers, eating up volumes of smaller, non-salable timber. Water-powered and steam-powered sawmills drove vertically mounted, straight-bladed crosscut saws (Muley Saws), producing flat slabs for flooring, doors, and siding. Even with intensive local use, forests were relatively sustainable resources until mining smelters and railroads came along, consuming huge quantities of wood.

With the coming of the railroad to Bull Creek in 1883, the demand for timber shot up. The expansion of railroads westward had created a newfound market for timber in southwestern Missouri,

Tie cutters. Retried from The State Historical Society of Missouri

as railroad ties and lumber for houses on the Kansas prairies. To supply this need, the government granted the Springfield and Southern Railroad several hundred acres of land on the ridge above the East Fork of Bull Creek. The timber was made available to "tie hackers," the hardy men who spent all day in the woods felling trees and then hacking them into the shape of railroad ties.

Along the rail line, towns like Oldfield and Chadwick grew up near Bull Creek, where hackers could load their ties onto railroad cars. Chadwick was at the end of the line, on a narrow ridge with steep slopes into the Bull Creek watershed to the west and Swan Creek to the east, both valleys producing thousands of ties. With the arrival of the railroad, businesses sprang up rapidly in Chadwick—stores, restaurants, saloons and the inevitable gambling houses and places of ill-repute. Like other boom towns of the era, Chadwick quickly became a rowdy, lawless place—a "Hell's Half Acre." A vigilante group known as the Baldknobbers formed in the area, further increasing the violence, ending in the 1887 murders of the Eden family a few miles north of Chadwick.

Above: Railroad ties stacked at the tie yard in Branson

Left: A tie hacker at work

The McCoy sawmill after the boiler exploded

With the availability of low-cost timber, workers began to pour into the area around Chadwick. Any able-bodied man could suddenly earn "cash money" from hacking and selling railroad ties, providing his family with clothing, farm implements and other goods. Hacked ties brought 55 to 65 cents apiece in the 1920s, with up to 85 cents paid for a premium tie. Many hackers harvested trees from unclaimed government land. Speculators had already purchased many areas of upland forests for the mining rights, creating conflicts with local tie hackers who had been harvesting that timber for free.

To make ties, hackers cut down trees with a crosscut saw and squared the sides of logs with a broadaxe. A good axman could produce fifteen to twenty ties a day. Two-horse teams and wagons hauled ties to the town of Oldfield, just up the line from Chadwick, for ten cents apiece, a trip that for some required fording Swan Creek five times. Sometimes haulers would have to unload half of the ties to make it up a steep hill in two trips, or else hitch up a second team.

Logging was an extremely dangerous occupation. In addition to falling trees, loggers risked injury from hacking ties and hauling logs. Ties were sometimes cut at mills, but more commonly mills produced boards for building houses and barns. Accidents at sawmills were common. Bull Creek resident J.J. McCoy was operating his steam powered sawmill in 1908 when a pressure relief valve failed. As he struggled to open it, the steam boiler exploded, scalding him severely. He died an excruciating four days later.

By 1930, most of the easily accessible timber had been cut. It would take many years for the timber to grow back to the size that it had been before logging began. The loss of trees may have caused some localized erosion from exposed soil, but what may have created more problems were the primitive roads or trails providing loggers access to usable stands of timber. Where heavily used trails crossed creeks and muddy drainageways, ruts could become deeply eroded by runoff. For milling, only the biggest and straightest trees were used. For this reason, damage to Bull Creek forests wasn't quite as severe as the "clear cutting" that dominated logging in other parts of the country.

After the logging boom, the rail line to Chadwick remained open for a while with passenger service. The "Chadwick Flyer," as it was sarcastically named, was famous for its slow pace and multiple stops. Locals joked that you couldn't grow anything beside the tracks because the slowly moving train blocked out the sun. The train's crew was rarely in a hurry. They would pause along the way to shoot quail or turkey or pick blackberries, leaving this bounty along the tracks to be picked up on the return trip. Rail service to Chadwick ended in 1934.

When the timber was gone, landowners held large acreages with almost no value. Large sections in the Bull Creek headwaters were owned by North Eastern Zinc Company, Hobart Lee Tie Company, or other private owners "back east." These speculators were clever marketers of cutover land, supported by exaggerated claims in prominent advertising. The countryside around Bull Creek was typically described as "not very rough or mountainous," or "undulating with some high plateaus," or "fertile virgin land containing limestone." Advertisements pushed land at $3.50 to $12 an acre in blocks of 80 to 1,000 acres. Having once supported trees of wonderful size, largely denuded hillsides were now subject to erosion.

By the end of the timber boom, many families had established deep roots in the Bull Creek area and were reluctant to leave. Tie hackers and mill workers settled into subsistence farming. They began seasonal burning of cutover woods to encourage fresh grass for grazing and to get rid of ticks and snakes. This practice suppressed the regrowth of timber and woody plants, further depleting the leaf litter and organic matter that formerly covered the forest floor. Attempts to grow crops on old forest land resulted in erosion and rapid runoff. With the leaf litter gone, there was less absorptive capacity and therefore diminished groundwater recharge. Many once-perennial springs dried up, depleting yet another resource for the settlers.

Where there are forests, of course, there are fires. For thousands of years, humans have created forest fires by accident or on purpose, shaping the Missouri landscape. While lightning sparks an occasional wildfire, human activity today causes over 90% of fires. Native Americans, we now know, used fires to their advantage. In 1750, Father Vivier described the effect of this burning:

> ...wherein trees are almost as thinly scattered as in our public promenades. This is partly due to the fact that the savages set fire to the prairies toward the end of autumn, when the grass is dry; the fire spreads everywhere and destroys most of the young trees.

Schoolcraft, trekking the Ozarks in 1818-19, described a landscape of prairies and "open" savannas and oak-pine forests. He did not realize that Native Americans had used fire as a management tool. Lacking ways of controlling the burn, Native Americans would allow wildfire to spread for miles, widely suppressing the growth of trees. Burning every few years allowed less time for brush and limbs to accumulate on the forest floor, making the fire burn less hot. Young trees and brush would be killed, but not the larger trees. The burning of open areas, especially the hilltops, had little or no effect on the stream banks and mature timber. Shrubs were removed, but grasses took their place. Overall, the ecosystem wasn't harmed.

With burning, over time forests transitioned to savannas, with grasses and bushes growing between widely spaced trees, giving the forest an "open" look and feel. Once established, this savanna landscape was routinely maintained by periodic burning. With long-established vegetation holding deep roots in the soil, erosion was minimal, even after successive burns. Open grasslands and savannas provided food for bison, elk, and deer. Openings in the forest also stimulated sun-loving fruits and berries.

Second growth timber in the Bull Creek watershed

Trees hold the history of fire within their trunks—a story that foresters can read. A fire of moderate intensity usually scars the cambium in that year's tree ring without killing the tree. These scars record the scope and even the direction of a fire. Dendrochronology, or tree-ring dating, can also note the exact calendar year of the event, in some cases as far back as the 1600s. Studies of dendrochronology have shown a link between human population density and the frequency and intensity of fires over the last four-hundred years. While sparsely populated areas had fires every ten to thirty years on average, more populated lands recorded fires every one to five years. As groups of Native Americans were driven westward and the population of fire-using Euro-American settlers expanded, fire frequency often increased.

Not until the first quarter of the twentieth century did the government step in to manage and protect Missouri's forests. The passage of the Weeks Act in 1911 allowed the federal government to purchase degraded forest lands for sustainable wood production as well as to protect watersheds and wildlife habitat. The Mark Twain National Forest was created, now encompassing over one million acres in Missouri. About a fifth of the land in the Bull Creek watershed is in the national forest.

In 1925, the Missouri Legislature appropriated $10,000 to create a Department of Forestry, but the governor vetoed the bill. The idea of protecting forests on state or private lands might have died there except for the creation of the Missouri Forestry Association, a coalition of private citizens and wood-using industries. The Association raised funds from public subscription and hired a forester for fire prevention education. He drove around in a Model-T nicknamed the "Showboat," hauling a trailer with a generator and motion picture projector, showing the film *Trees of Righteousness*. In the days before widespread electrification, the mere presence of a movie in the woods brought out curious crowds.

Today, the Missouri Department of Conservation, created in 1937, also works toward better management of Missouri's private forests. Enhanced fire control measures in the 1960s, along with robust reforestation programs promoted by the Conservation Department, has led to a significant regrowth of forests, both private and public. Accessible, publicly owned, professionally managed forest land has created increased opportunities for hunting, hiking, camping and, in some designated areas, off-road vehicle recreation.

Forest management has advanced over the years. We have rediscovered the value of fire as a tool for maintaining high-quality prairies, glades and even woodlands. Prescribed burns (a controlled fire with a defined purpose) can foster improved wildlife habitat, plant diversity, and even timber production. With controlled burns, brush and understory plants can be temporarily removed, leaving more growing room for large and healthy trees. With increased space and light, the remaining trees will grow larger and faster. Increased light penetration between trees also benefits wildlife. Grasses and bushes will grow in sunlit openings, providing important food sources and cover for animals. Although an important food source, acorns are only around for a few months in the late fall. Low-growing vegetation between trees provides food and cover year-round.

Both private and government forest owners now have better techniques available for managing forest lands, including practices that improve the ecological values of forests. For example, it is especially important to leave a buffer zone of trees along streambanks, with at least a one-hundred-foot buffer recommended. In upland areas, management may include the removal of some species or individual trees, allowing the remaining trees to grow larger. In the case of oaks, these larger trees will produce abundant acorns to feed turkey and deer, and to fatten up bears in preparation for their winter hibernation. Some old trees are left as dens and nesting sites for wildlife, as well as providing food sources for birds such as woodpeckers. As our tools improve, so does the value of our forests and woodlands. The view of forest trees rising above Bull Creek is inspiring, and a welcome reminder that a healthy forest provides myriad benefits to both man and beast.

CHAPTER 7

FLOATING BULL CREEK

Floating is the best way to get to know a stream. Floaters get a front row seat, getting an up-close and personal peek into the stream's complex personality unfolding mile after mile. Curiosities appear at every turn—a heron lifting, grumbling, pterodactyl-like from the shallows; or a swooping, scolding kingfisher; or a dinner-plate-sized turtle tip-toeing along the bottom. The lush, heavily vegetated area near the stream is rich habitat, home to a wide variety of wildlife. Human bonds with the natural world can be sensed strongly here.

Streams like Bull Creek are particularly well suited to floating. The current is swift and steady, but not overly dangerous at reasonable flow levels, providing a respectable pace and some exciting riffles and fast chutes. Although Bull Creek is normally a mild-mannered stream, its temperament can change dramatically at different levels of flow. When in flood, Bull Creek is dangerous for canoeing or kayaking. On the other hand, if the flow is too low, exposed rocks and ledges will cause the boat to scrape or get hung up, requiring the floater to get out and drag the boat through. Medium levels are the best, when the water is up slightly but still clear, and most of the rocks are safely covered.

People who float over many seasons and years will notice other changes. For example, we now have very good records on streamflow. It is easy to find out if there were more or bigger floods on Bull Creek this year than last year, or ten years ago. But if there are, how do we explain these changes? Are we seeing long-term changes in response to new weather patterns, an anticipated result of climate change, or natural year-to-year variability? The emerging pattern seems to be bigger floods over the last several years than in the previous hundred, since flow records have been kept.

The lengths and durations of low flows may also be changing. Like many Ozark streams, Bull Creek is sustained during dry times by springs. Were it not for springs, much of Bull Creek would be bone dry during long, hot summers. Life in the stream is dependent on this sustained input of cool spring water for survival. But is there less spring flow than there used to be? It's an important question, but one that is difficult to answer definitively without good background data. Anecdotal information about changes in spring flow may suggest trends, but it is hard to say with certainty if these are long-term changes.

Noppadol Paothong photographing in the swimmin› hole

Gravel bars, which are plentiful on Bull Creek, also change dramatically from year to year. Each new flood moves thousands of tons of gravel, depositing it downstream in new gravel bars or removing it from older bars. The bigger the flood, the more gravel that is moved. Streamside landowners often complain about all the new gravel piling up on their properties or smothering their farm fields. Very often, they blame the increase in gravel or upstream logging or development. That may be true, but are we witnessing the results from human disturbances that happened long ago, or more recently? Or, is this another phase in an ongoing cycle of natural change? Or is it the result of more and bigger floods attributable to climate change?

Jud Whitlinger baiting up on the gravel bar

A few years ago, float trips on Bull Creek usually began at the Goodnight Hollow Bridge, a convenient put-in off Round-Mountain Road. There was an old low-water concrete bridge there where children constructed rock dams and people swam in the summertime. But the old bridge was recently replaced by a nondescript high bridge. Public access to the river was cut off by fencing and "no trespassing" signs, as it has been at many other Ozark streams. But in the long term, denying public access could be bad for the stream. People tend to care about what they know and love. By denying access, we may be silencing a constituency that otherwise would take a great interest in the welfare of Bull Creek.

The first stretch of creek below the new high bridge is barely a foot deep, the current gently slipping along. Sitting back in the boat, the floater can relax and enjoy a mellow ride down the leafy corridor. There is little fishable water up here, just a few scour holes under boulders and pockets beneath snags. About a mile downstream, the river pushes into a deep hole beneath a bluff. This is the first of many deep fishing and swimming holes in Bull Creek. These deep blue pools of slower water are the domain of the most quintessential finny citizens of the creek, smallmouth bass, a favorite of Bull Creek anglers.

In the late spring or summer, smallmouth may have to be enticed from beneath logs or ledges. Deep holes hold the meatiest fish, while smaller ones are normally caught in riffles or shallow water. But even a twelve-inch fish puts up a good fight, especially when cutting against the vigorous current. Bass can be caught on most float fishing trips, but not always big ones. When any given trip is unproductive, it is natural to ask: Is fishing getting worse? If so, why? Is the river being overfished, or is some other factor at work? Changes in the river from season to season are natural, but are humans affecting those changes?

If the weather is hot, the floater may be more inclined to swim in the holes than fish them. A diver can get very close to fish, squaring off with staring red eyes and gulping jaws. Smallmouth are primarily sight feeders, like us, so the clear water of Ozark streams is prime habitat for them. They prefer cool water, too. Studies have shown that smallmouth bass grow faster in spring-fed streams like this than in northern waters, since spring-influenced streams stay at a more constant temperature year-round, inducing continuous growth. Recently, the summers have been extremely hot. Average temperatures are on the rise. How might this affect smallmouth bass, who are adapted to life in cooler water? Will anglers eventually have to travel farther north to catch them?

Fishing on Bull Creek, like most streams, is often hit or miss, to some extent. The time right after a flood is not a good time to fish. Bass feed like crazy when the water is rising right after the start of a rainstorm, when thousands of bugs and worms are being flushed into the stream with runoff. The fish no doubt get their fill at these times. Then, it's time for them to just wait out the high water, staying out of the current in quiet eddies, digesting all that protein. But huge floods, especially several floods in quick succession, can drastically reduce fish populations, as deep holes get filled with gravel or boulders serving as nesting areas get moved downstream. In recent years, we have seen some of the largest floods on record. Will fish populations be able to withstand such big flood events?

Copperhead, a venomous snake common in the Bull Creek area. The attractive skin pattern makes it very hard to see in the leaves

Wildlife is usually abundant near the stream. Animals come to drink, or gather food, or make homes along streambanks. Kingfishers scold loudly as they cross the stream in swooping flight. Huge old hollow trees on streambanks provide nesting sites for owls and wood ducks, or den sites for raccoons and opossums. Woodpeckers drum on dead snags, looking for insects, and black snakes slither up trees to raid bird nests. Beaver dens can sometimes be spotted at the base of mud banks, and in winter, it is common to see minks scampering along rocky banks. If the floater is quiet, he or she will be rewarded with glimpses of deer or otters, or even an occasional bobcat.

Most of the wildlife along Bull Creek is harmless, but some can be annoying. Case in point: raccoons seem to be getting more aggressive, especially on popular float streams. They have learned

Bridge on Bull Creek at low water. Notice water is flowing over the bridge rather than through the pipes.

to open latches on coolers and chew into plastic gear bags. Black bears, too, are a potential nuisance. They are extending their range. It's still rare to spot them on float trips, but these days, it is not a bad idea to hang a "bear bag" on overnight trips just in case, to discourage nighttime raiders of all kinds.

Some wildlife along the stream can be dangerous, even lethal. Along the banks of Bull Creek, especially near dusk, it's a good idea to watch where you step. Snakes are common in summer, including venomous copperheads, which can be hard to spot, coiled as they are in weedy or leafy areas.

The biggest hazard in Bull Creek is man-made structures, especially bridges. Like other streams, Bull Creek's "personality" is highly dependent on water levels. An easy float at low levels becomes much pushier and trickier at higher flows, when logs, limbs and structures increasingly impede the rushing water. In Bull Creek, and across the state, the number of recreational boaters has increased, especially since the outbreak of COVID-19. Over the last few years, thousands of kayaks have flown off the shelves

New bridge at the same site. Raised span allows fish to swim upstream.

of outdoor stores. A rise in boating accidents and drownings could result from more people getting on the rivers, many of them with little floating experience. Many of them do not have adequate rescue skills, nor do they have any thoughts of matching their skill levels with the level of river flow.

Unfortunately, Bull Creek has claimed the lives of floaters. Three boaters have drowned on the stream since 2019, all at or near the same spot. In 2017, a riverside landowner built a concrete low-water bridge across the stream to connect his properties on either side. At low water, boaters were forced to stop at the bridge and drag their boats over it. Pipe culverts had been built into the bridge to allow water to flow under it, but these openings quickly became plugged with gravel and brush, forcing water to flow over the top. At low water, the bridge wasn't a big deal—one simply slid their boat over the smooth, algae-lubricated surface of the bridge. But at high water, it became a very big deal.

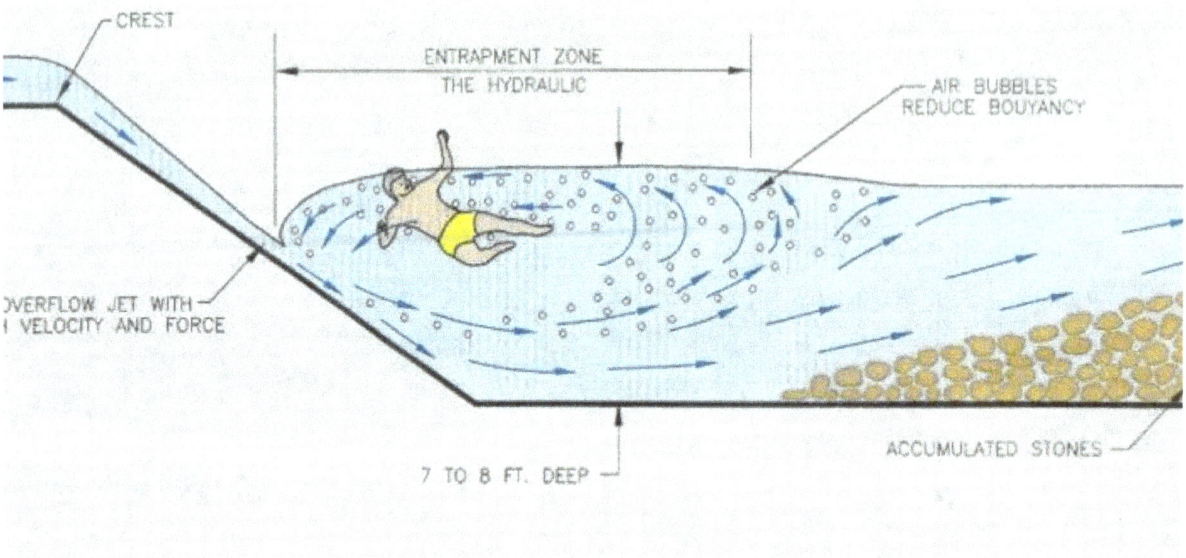

Dangerous hydraulic is created where water flows over a dam or bridge

The problem is that at higher levels, water flowing over the top of the bridge forms what is known in river runners' parlance as the "hydraulic," or "hole." The flowing mass of water coursing over the bridge plunges to the bottom on the downstream side, pushing the water ahead of it downward. Water near the surface, just below the drop, actually flows upstream to fill in the valley, the "hole," created by the downwardly plunging flow (see illustration, above). The result is that objects that float, like a log, or a barrel, or a person in a life jacket, can become trapped in the hole's mean embrace. Holes produced naturally, as by boulders, are usually irregular in shape, often providing for escape at the ragged edge. But if it's a smooth hydraulic, like one formed by a dam or bridge, a person can get hopelessly stuck.

For boaters who aren't prepared for what the water will do to them, the violence of a large hydraulic will almost invariably flip the boat. With luck, the person, now in the water, will flush out downstream. If wearing a life jacket, they can swim to shore or, if someone is nearby, be rescued. But in higher water, and particularly with no life jacket, a swimmer may be pounded by rocks and waves and have a long swim or struggle to get to shore. A long, panicky swim in cold water and powerful currents while trying to keep one's head above water quickly saps the energy. Hypothermia, at least, and often drowning can be the result.

Center Road Bridge over Bull Creek. Notice that the "drop" is too high to let fish swim upstream.

The three Bull Creek boaters drowned between 2019 and 2020, after the installation of the bridge, and all at high or medium-high water levels. At least two people, one of whom drowned, were not aware that the bridge had been installed since they last floated Bull Creek. After this latest drowning, in June 2020, the Army Corps of Engineers inspected the bridge and determined that it had not been properly constructed. The Corps directed the owner to tear out the structure and replace it with a new one in compliance with permit requirements. The owner did just that. But even with larger openings beneath it, the new bridge is still so low to the water that a canoe can't pass under it at most levels. This, plus the closing of accesses, has constrained the floating opportunities on Bull Creek.

Man-made structures in streams are more than nuisances, or eyesores, or even hazards for drivers, boaters or swimmers. The Missouri Department of Conservation has identified dozens of low-water bridges and stream crossings in southwest Missouri that are barriers to the migration of fish and other aquatic organisms. Fish have trouble swimming upstream through tubes of high-velocity water under bridges—it's like fighting against a water jet. Or if there is a ledge or drop below the bridge to get over, no matter how small, their upstream progress may be blocked. Limited funds are available from public agencies to replace problem structures with raised span bridges that not only allow for fish passage but are safer for vehicles at higher water levels. In the future we need to design and build stream crossings with nature in mind, in addition to human safety.

More and larger floods pose increasing dangers for boaters, riverside landowners or those crossing swollen streams in vehicles. Scientists have been warning us for years that these changes could very well be related to changing climatic conditions, which in turn result from a warming planet. Streams have always flooded, and many organisms time their reproductive cycles to a flooding event. But stream ecosystems that have evolved to tolerate and even anticipate the inevitable floods of the past may be hard pressed to survive ever more frequent and bigger events. The big flood on the North Fork in May 2017, for example, vastly reduced the trout population in this blue-ribbon fishing stream. People working in river and streambank restoration worry that measures designed to deal with ordinary floods, like tree plantings to hold streambanks in place, won't be able to withstand super-sized events and will be ripped out.

In the Ozarks, we have been isolated somewhat from the major disasters spreading in other parts of the country—the extreme droughts and big fires in the West, the devastating hurricanes along the eastern coasts. But things seem to be changing here as well—bigger, more frequent floods, for example. Other environmental problems, which might be linked to climate change, are more subtle, but still worrisome—the depletion of amphibians, insects, and songbirds, for example, and more invasive species stealing habitat from the natives. It has become rare to hear the exuberant calls of a whip-poor-will on Ozark streams. There seems to be fewer songbirds and deep-throated jug-o-rum bullfrogs or twanging green frogs in the spring and summer. This acoustic vacuum is troubling to river runners who have become accustomed to, and comforted by, the melodious sounds of wild nature.

Like other environmental stressors, climate change is a mixed bag. For some creatures, like armadillos, the warming planet may be helpful, allowing them to continue their march northward. But in new northern homes, they will face new competition from the species that already live there. The picture of exactly what changes will occur with the changing climate is at this point very fuzzy. Some species have the ability to move with the changing times—others don't. The more resilient species may be able to move and prosper in new territory, while others will continue dropping in numbers until they finally go extinct. And for those that move, will they outcompete and displace some of the species in their new homelands? All these questions remain unanswered. And finding answers takes time—time that we may not have before changes are already taking place. Whatever the final questions are, the answers will certainly affect Bull Creek and its watershed.

Change is often disruptive and can be very troubling. It is difficult to imagine the rich fabric that is Bull Creek becoming faded, frayed, and torn. Still, this little Ozark gem remains a beautiful, thriving place. The water still runs clear and cold as it pulls the canoe silently into a gentle riffle, gliding the boater past a green-backed heron hunkered at the water's edge. It is comforting to know that a place like this can still exist so close to home. On the river, there is no need to be profound, or witty. It is a place to relax, breathe healthy air tinged with river aromas, but robust and clean—to lean back, watching a red-tailed hawk circling on the thermals, high overhead. There is time to enjoy the view, and the surrounding scents and sounds. Away from the maddening traffic and hectic schedules, the boater is no longer on city time—he or she is in resonance with an old friend—one that fits and feels like a well-worn sandal—river time.

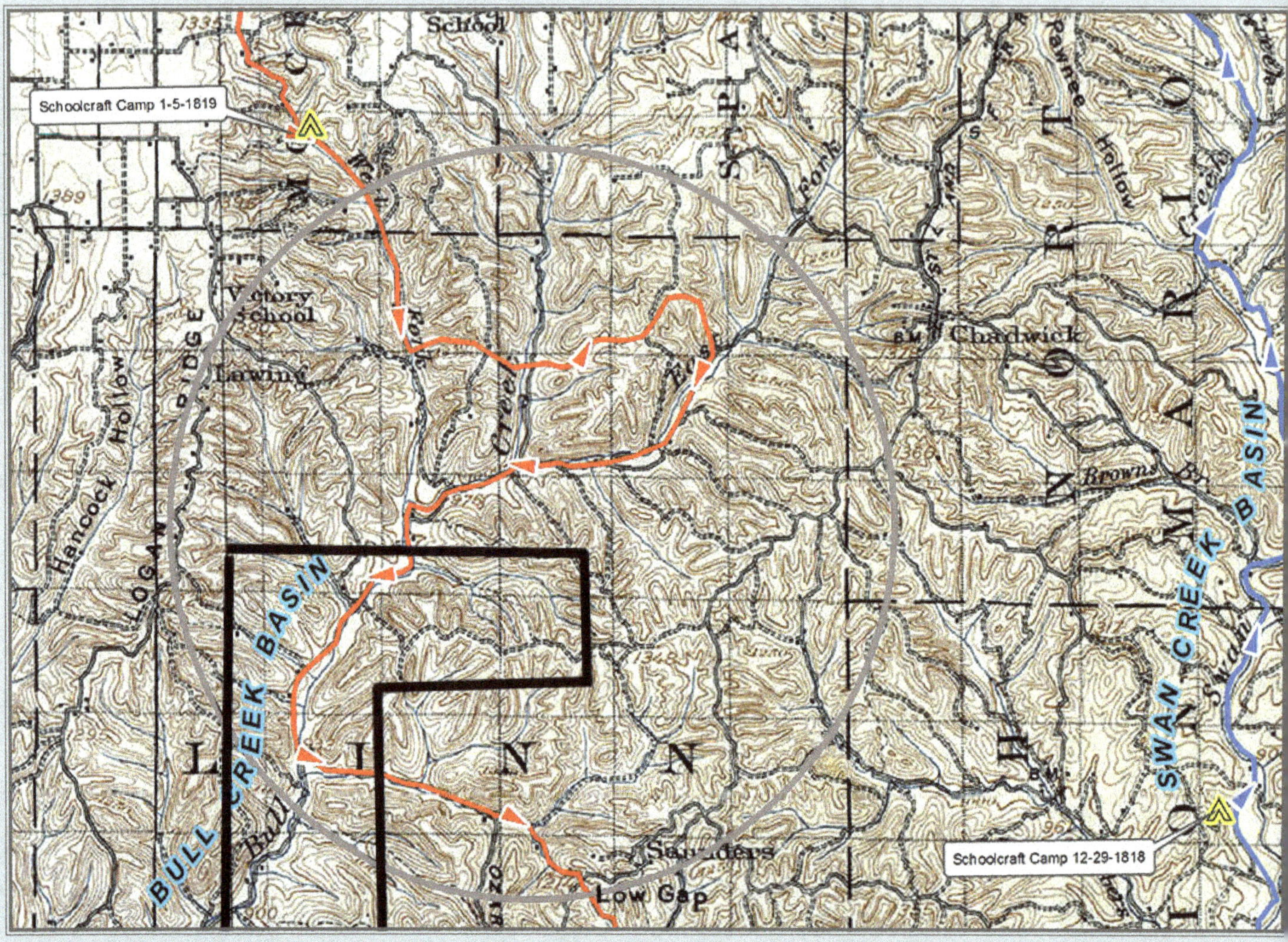

Schoolcraft's route through the Bull Creek valley. (Kipfer's property shown by the bold outline.) Map courtesy of Curtis Copeland

CHAPTER 8

THE NEW WAVE

Over the past two-hundred years, the Bull Creek valley has gone from nearly untrammeled wilderness to the comfort, security and convenience of suburban real estate. Schoolcraft, traveling through the valley in early 1819, mentioned no cabins or settlements near Bull Creek. In fact, he said the area was even avoided by hunters because of its high ridges and confusing terrain. The first noteworthy communication link through the valley was the Ozark Mail Trace, blazed in the 1830s. This rough road followed part of Schoolcraft's route after he became lost and wandered into the Bull Creek Valley, then escaped the valley by ascending Peckout Hollow, up and over otherwise unbroken ridges into the valley of Swan Creek.

Initially, the White River was the main transportation artery into the Bull Creek country. From a landing on the White River near the future town of Forsyth, mail and trade goods would be carried on the Mail Trace Road northward to the commercial hub of Springfield. Before much longer, that road would also link Forsyth to Carrolton, the farthest west post office in Arkansas Territory. The Mail Trace is reminiscent of the only kind of road that once existed in this part of the country—slow, rocky and rough going.

In the early days, there were no criteria or standards for what constituted a public road. In 1837, the law defined a public road as one, "at least 30 feet wide, with at least 15 feet of it being cleared of all stumps, trees, grubs and other obstructions." After Henry Ford introduced his Model T in 1907, a series of new state laws sought to improve roads at the county level. State funds became available for county road maintenance in 1909. But for many years after that, most of the roads in rugged sections of the Ozarks, like Bull Creek, were rutted and steep in many sections, and muddy or frequently flooded in others—in other words, often unpassable.

The Mail Trace (also known as the Springfield-Forsyth Road) was still largely unimproved as late as 1915, except for occasional leveling performed by dragging a log perpendicular along the trail. On it and other roads, river gravel was spread to keep tires out of the sticky red mud or slimy gumbo. The easiest place to build roads was in the valley bottoms, where stream gravel for re-surfacing was

Remains of Mail Trace through the Kipfer property

near at hand. Bridges were nearly nonexistent until the twentieth century, with most roads crossing creeks at fords.

Electricity arrived a little later than improved roads. Powersite Dam on the White River was completed in 1914, bringing electric power to the region. A power line was extended northward from the powerhouse and dam near Forsyth, through the Bull Creek valley, to the cities of Ozark and Springfield. Landowners along the line gladly gave right of way to the power company, assuming they would be first in line for electrical service. But often, connecting them would take quite a while, as the cost of running ancillary power lines to scattered individual homes in the valley was at first prohibitive.

Even with access to electrical service, its cost kept homeowner utilization rates low. This would change in 1935 with the creation of the Rural Electrification Administration (REA), which provided incentives to form electrical cooperatives, bringing lower cost power to sparsely settled sections. Soon, houses in the hinterlands had lights, water pumps and labor-saving appliances. Telephone service reached the valley a little later still.

Power pole that carries electricity from Powersite Dam to Ozark and Springfield

With the end of World War II, jobs in war industries collapsed, as did opportunities in lead and zinc mining. Many families once supported by these industries returned to the hills and hollers of their predecessors. By then, rural life had become a little easier. Electricity made it possible to have potable water on the ridgetops, where drilled wells could be equipped with submersible pumps. All-weather roads provided easier access to cities, where new jobs had opened. Daily travel to stores and restaurants became possible.

While early settlers were attracted by strong family and social ties, today's residents are typically commuters looking for cheap land. They enjoy immersion in the rugged outdoors and proximity to "nature." With Bull Creek located so close to urban areas, suburban residents can make a decent

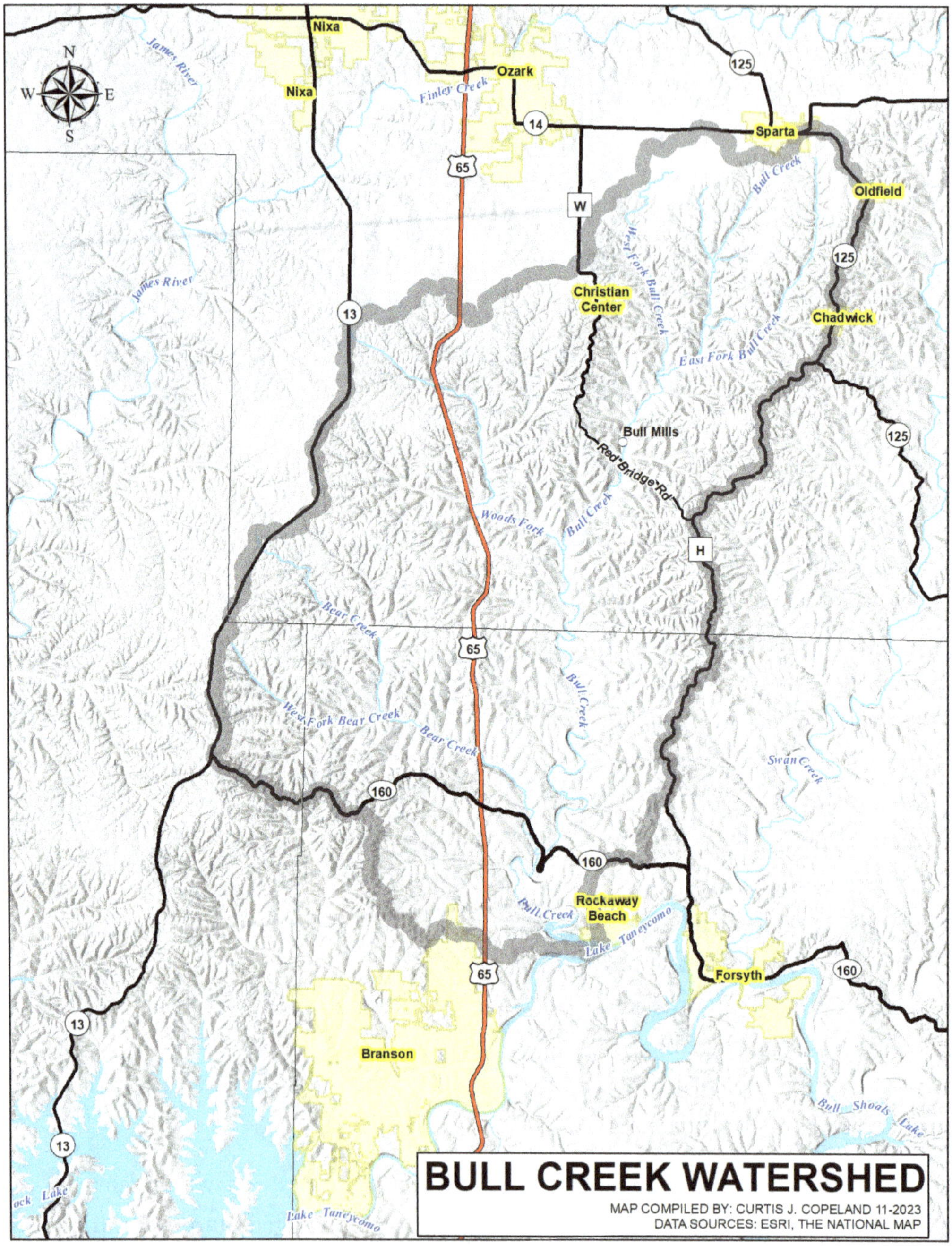

Map of the Bull Creek Watershed. Bull Creek watershed denoted by bold gray line.

living at a job in the city while holding onto a rural way of life. This has resulted in a tremendous growth of rural residential housing, spilling over from the rapidly urbanizing areas of Springfield, Ozark, Nixa and Branson. The northern two-thirds of the Bull Creek watershed lie in Christian County, which from 2000 to 2010 was the fastest growing county in Missouri and one of the fastest in the nation. There are lots of new subdivisions on the outskirts of Ozark and a proliferation of houses along rural routes.

Transplants from urban areas introduced a new dynamic to the quiet countryside. They want that idyllic country life, yet still expect all the urban amenities. They work to make their rural property resemble a tidy slice of suburbia—with mowed lawns and non-native ornamental bushes and flower beds and statuary. Like their urban counterparts, they want a few trees, yet don't want leaves piling up in their yards. They want birds, but don't understand that manicured turf offers meager habitat and scant food for them.

While it looks nice, the immaculate suburban home with its extensive lawn neatly trimmed, fertilized, and dandelion-free may present a threat to water quality and watershed health. How many pounds of fertilizer and pesticides are being applied to that seldom-used, super-sized yard? How much of this is staying put, and how much is leaving the site in runoff? How much groundwater is being pumped to fill the swimming pool or irrigate the landscape? How much pollution-generating energy is being used by the lawnmower, or the SUV commuting daily to an urban center?

Most homes in the Bull Creek watershed are beyond the reach of municipal sewer systems. Counties are responsible for inspecting and permitting new septic tanks, now more properly referred to as onsite wastewater systems. The rules are better than they used to be, but the sheer numbers of new installations create concerns about whether ground and surface waters are being adequately protected. The most scenic, desirable building sites are in the rugged hills, where steep slopes and thin, rocky soils present difficult conditions for the proper functioning of onsite wastewater systems. The cumulative effect of so many individual systems may be a gradual deterioration of water quality in Bull Creek.

In a way, though, it's a good thing that sewers haven't reached the Bull Creek watershed. Sewers are only cost-effective in areas with hundreds of closely spaced homes, where sufficient revenue can be generated through sewer bills to support expensive public treatment systems. In these highly urbanized areas, acres of impervious surfaces such as rooftops, roads, driveways, and parking lots prevent the infiltration of rainwater. Instead, these hard surfaces quickly funnel stormwater runoff and

Church in Chadwick

its considerable load of pollutants to the nearest stream. Water quality may not be helped if problems from malfunctioning home sewage systems are traded for problems created by polluted runoff.

The rapidly urbanizing areas of Branson and Springfield exert profound influences on the Bull Creek watershed, sandwiched between them. Branson and Table Rock Lake to the south funnel money toward tourism, recreation and entertainment, while medical services, manufacturing and small business opportunities attract capital northward toward Springfield. Many people connected with these industrial and commercial enterprises, seeking closer connections to nature, perhaps, or looking for seclusion, have found exclusive homesites in the Bull Creek watershed. Many of these homes are noteworthy—some are downright astounding.

Sunrise over the Bull Creek valley

Recently, new kinds of commercial enterprises have sprung up in the Bull Creek watershed near U.S. 65, hoping to cash in on the vigorous Springfield to Branson trade. There is the National Tiger Sanctuary along Bear Creek, a "rescue organization that provides a permanent home for exotic and domestic animals." Founded in 2000, the facility houses big cats—leopards, pumas, lions, and tigers. Another novel attempt is the Copper Run Distillery, the "first legal distillery in the Ozark Mountains since Prohibition ended in 1933." This small-batch craft distillery, reminiscent of hillbilly lawbreakers, is "heavily influenced" by "pure limestone-rich water."

Bear Creek Winery, near Copper Run, hosts local musicians, prepares "rustic pizza" and features "cozy heated wine huts," resembling tiny greenhouses, overlooking the Bear Creek valley. Among

New county bridge over Bull Creek

their wine offerings is a white called "Hillbilly Persuasion." One of the most prominent attractions is just a few miles down the road—a very tall white cross on a bluff overlooking Highway 65 painted with large water drops, or tears. A distillery, a winery, a tiger sanctuary and a huge cross—all in a few-mile stretch within the Bull Creek Watershed. About the only thing to be said about this rather bizarre juxtaposition is that it highlights the stark differences in American visions of entertainment and lifestyle.

Farther south on U.S. 65, just north of the city of Branson, U.S. 160 turns to the east, toward the lower end of Bull Creek. This is Taney County, home of "The Shepherd of the Hills," the "Bald-Knobbers" and "Silver Dollar City," and containing the lower third of the Bull Creek watershed. After many curves and undulations, Highway 160 approaches Lake Taneycomo (named for Taney County, Missouri) at the mouth of Bull Creek. Here are many older resorts, platted villages, fishing accesses and newer residential properties, intermingling low-income with the well-to-do.

The village of Bull Creek, near the mouth of Bull Creek and partly in its floodplain, consists almost entirely of mobile homes, many of them abandoned. In nearby Merriam Woods, many homes are dilapidated, the formerly paved streets now rough and rutted. Low tax bases in both communities have made it difficult to provide timely upgrades for wastewater systems. Nutrients from poorly treated sewage have caused profuse growths of water plants in the Bull Creek arm of Lake Taneycomo, limiting recreational uses. The village of Merriam Woods finally hooked to a modern wastewater treatment plant at Rockaway Beach, near the mouth of Bull Creek, solving at least the immediate sewage problem.

Rockaway Beach was a popular resort area by the late 1950s. People swam, boated, or fished in the relatively warm, uppermost reaches of Bull Shoals Lake. But when a hydropower dam was built upstream to form Table Rock Lake, cold water flowed from the bottom of the reservoir, filling the channel of the old White River, now called Lake Taneycomo. Cold water sustained trout fishing but made the "beach" at Rockaway much less enticing for swimmers. Visitation dwindled. As Branson boomed, Rockaway deteriorated.

Many portions of the Bull Creek watershed, especially near urban areas or along highways and county roads, are either poised for development or already in a patchwork of commercial and residential uses. What does this mean for the future of Bull Creek? Will more houses, convenience stores, gas stations, restaurants, and resorts mean more polluted runoff and shallow groundwater contamination from onsite wastewater systems? Will all this development ultimately affect the clarity and cleanliness of Bull Creek, or degrade aquatic habitats and decimate the fishing?

It is hard to say, of course. But one thing is certain. Growth will almost certainly continue in the areas around Springfield, Ozark, and Branson. And with climate change upon us, more and more people will be eyeing the Ozarks, with its abundance of clean water resources, relatively low cost of living and absence of disasters like hurricanes, earthquakes, and massive forest fires. Climate "refugees" could well go from a trickle to a torrent as large-scale atmospheric disturbances become more intense and widespread.

On the positive side, people who move here because of problems elsewhere might be a bit more motivated to protect and defend their new homes. Historically, Ozarkers have been wary and mistrustful of outsiders. But those transplants have often become the most vigorous defenders of their newly found, relatively unspoiled homeplaces. They may be less inclined than the "natives" to take the high-quality of their environment for granted.

Those of us who live here, whether natives or newcomers, have an obligation to take care of those things that make Bull Creek, and the Ozarks, such a great place to live. To do that, we will have to foster honest and meaningful communication with people who don't agree with us about many things. Somehow, we must bridge the great divides—socioeconomic, ideological, political—the continental divides, it would seem, that increasingly separate our diversely populated landmass. We must convince developers, homeowners, and businesses that keeping Bull Creek healthy and alive is in all our best interests. We can't stand by and helplessly watch it become fouled and degraded.

There is just too much at stake to let that happen. Bull Creek, so near to urban areas, is an economic as well as environmental asset. People want to own land and live near a clean stream surrounded by natural beauty. These attributes keep real estate values high. The stream also provides outstanding recreational opportunities—a place to swim and snorkel in clean, crystal-clear, spring-cooled water, or take a pleasant float trip by canoe or kayak. Most importantly, Bull Creek is a biological treasure trove, providing refuge for diverse organisms that have great value in their own right—the pollinating bees and butterflies, the fish we like to catch, the birds that so dazzle and enliven our outdoor world. If Bull Creek as we know it today is wiped out, we will all be much impoverished. It would be a sad day indeed if such a wonderful resource is tarnished or lost.

CHAPTER 9

THE FUTURE OF BULL CREEK

With Bull Creek—as with most things in life—we cannot know what the future will hold. Some of our region's finest streams have been recognized by law as outstanding natural resources and therefore achieved a degree of protection. But most, like Bull Creek, have no level of oversight other than that provided in state or county laws dealing with development or pollution. Many organizations and agencies are working on behalf of streams and environmental quality in general, including efforts to manage watersheds, regulate pollution sources, and permanently protect land through instruments such as conservation easements. Obviously, these efforts must be grounded in public education.

Organizations like Master Naturalists, Audubon, the Ozark Society, Ozark Land Trust and Ozark Greenways are busy spreading the gospel about the link between human and ecological health and the important ecosystem services of native landscapes. Organizations like the Watershed Committee of the Ozarks and James River Basin Partnership remind us of the value of clean water and the need to use better practices on our farms or suburban or urban properties. In Springfield, a program called Yard Ethic teaches homeowners the value of native plants in managing their yards, their miniature "biotic communities," in ways that prevent water and air pollution while helping bees and other beneficial insects and songbirds.

In the long run, will these efforts and programs be enough to save Bull Creek from an unwanted fate? Probably not. If the creek is to be saved, it will require more attention from those who directly care about it, especially those living on its banks or in its watershed. Luckily, there already exists in the watershed a core group of people with strong conservation ethics. One of these is Jennifer Ailor. She and her husband, Don, bought property in Christian County in the late 1980s, where they built a beautiful home overlooking the Bull Creek valley. At the time, there were no other homes in sight. But within a few years, houses began to spring up all around them. Since then, the pace of development has never let up.

Jennifer Ailor and her dog, Bella, on the driveway to her house

The back of Jennifer's house abuts a rocky bluff, where she grows patches of wildflowers in the comfy microclimate. "Things blossom down here way before they do at the top of the hill," she noted. Other than the immediate area around the house, trees on the 91-acre tract were spared during construction.

"What aggravates me most," she said, eyes flashing, "is when people buy a piece of land with woods on it and then cut down all the trees—they want to see out or have a nice lawn. Then why did they pick that kind of property in the first place?"

It's a good question.

Many conservation-minded people acquired property in the Bull Creek watershed because of its quiet, rural character. Curtis Copeland and his wife, Crystal, bought property in the watershed near Branson, where Curtis is the GIS Coordinator, to be near the wild; to obtain "quick, easy access to remoteness." They found a comfortable home amidst a tight-knit community of neighbors. We "keep in contact," Curtis said.

Curtis and Crystal Copeland with their children

Larry and Judy Oliver at their home on the banks of Bull Creek

"We know each other's kids; know each other's dogs." A feeling of wildness is reinforced by the surrounding deep valleys and wooded hills. "We still hear the coyotes at night; we still hear the hoot owls."

Fred McQueary bought land along Bull Creek in the late 1980s, originally to provide a place for hunting. He liked the fact that it was "geographically isolated," and that the creek formed a barrier against trespassing. But the property also became a family refuge and retreat where his kids swam and played in the creek and caught "critters." As he steadily worked to improve the land—through prescribed burns, glade and savanna restoration and the establishment of warm season grasses—he also improved soil health and vegetative cover, in the process safe-guarding the high quality of Bull Creek. Today, he has permanently protected his restoration investments through a conservation easement.

Larry and Judy Oliver live on Bull Creek at the end of a rough,

Bull Creek painting by Jane Troup

Upper Bull Creek from the Oliver's house

Magnificent spring day on Bull Creek. Photo by Kyle Kosovich

partially paved dead-end road. What attracted them to this remote, rather hard-to-reach place? It is surrounded by national forest, they said, which decreases the likelihood of any new developments going in nearby. But Larry suggested that their creekside neighborhood was also unique.

"From our southern border to Center Road, for about eight miles, there is a community of landowners who are essentially all environmentalists, all conservationists,"he said.

These ethically driven neighbors connect through a web of shared values—of outlooks and approaches that seek to protect the Earth, our life support system, rather than degrade or destroy it. It is a good thing that these people have taken up residence along Bull Creek. It bodes well for the future of the creek. But they also have help from nature. Upstream from them are thousands of acres of forest, holding soils on steep hills and streambanks, churning out oxygen, providing food and habitats—an expanse of forested land that Curtis Copeland suggests may be the "saving grace" of the Bull Creek watershed.

If those forests can be protected—by using management measures that sustain economically viable forestry while supporting the myriad ecological services provided by healthy forests; if farmers will continue to improve grazing practices and provide buffers along streams to protect water quality; if homeowners will maintain septic systems by periodic pumping and prevent the runoff of topsoil and reduce or eliminate pesticide and fertilizer use; if boaters, fishermen and hunters will abide by game regulations, respect private property, and continue picking up trash from thoughtless visitors; if landowners along the creek will leave healthy vegetation in riparian zones or plant wide buffer strips, or protect sensitive lands in perpetuity through conservation easements—it is indeed a big IF—but if so, the future looks bright.

Bull Creek is a wonderful resource deserving of protection. But don't just take our word for it. Find your own connections to Bull Creek—go swimming in one of its emerald pools in the bathing glow of a summer evening; or take a float trip down its chlorophyll-infused, amazingly alive, bird-filled corridor on a bud-bursting day in May; or recline on one of its clean gravel bars under the glorious scarlets and golds of fall, simultaneously warmed by the sun and cooled by the breeze; or wrestle a rowdy smallmouth from beneath one of its deep, secret limestone ledges. If you will do just one or more of these things, we believe that you, like us, will fall in l ove with this idea that is Bull Creek.

THE AUTHORS

Loring Bullard (left) and Bob Kipfer (right), live in Springfield with their wives, Beth and Barb, respectively. In this photo, they are awaiting transportation to their favorite watershed.

Endnotes

1. Milton Rafferty, *Rude Pursuits and Rugged Peaks: Schoolcraft's Ozark Journal 1818-1819,* University of Arkansas Press, 1996. Rafferty has done a great job of placing Schoolcraft's journal notes in a modern context. A map in the book's Introduction locates Schoolcraft's approximate route and camping sites.

2. Ibid., p. 43.

3. Frances Backhouse, *Once They were Hats: In Search of the Mighty Beaver*, ECW Press, 2015. Recent publications emphasize the ecological importance of beaver in the landscape, and how their nearly total elimination in parts of the country drastically affected hydrology and habitat conditions.

4. North American Wildlife and Habitat website: https://www.northamericanwildlifeandhabitat.com/missouri-black-bear-season-2021.html . This website has information about bear sightings, bear hunting seasons, and the "history" of black bears in Missouri.

5. Joel M. Vance, "Bears in Missouri's Ozarks Growl with Arkansas Accent," *White River Valley Historical Quarterly*, Vol. 4, No. 10, Winter 1972-73. Vance, who retired from the Missouri Department of Conservation, is an accomplished outdoor writer who uses humor to get his points across. Some of his book titles include *Grandma and the Black Bear; Tails I Lose: Coping with Bird Dogs;* and, *Save the Last Dance: A Story of North American Grassland Grouse.*

6. Ibid., p 6.

7. Missouri Department of Conservation, "American Black Bear Field Guide," website: https://mdc.mo.gov/discover-nature/field-guide/american-black-bear. This website has information on the life cycles of bears, ecosystem connections, human connections, and bear habitat and conservation.

8. Missouri Department of Conservation, "Missouri Black Bear Project," website: https://research.mdc.mo.gov/project/missouri-black-bear-project. This website provides updates on bear tracking, footage from trail cameras, and maps, photos and videos from bear research projects.

9. American Black Bear Field Guide, website.

10. Linda Hall Library website: lindahall.org/about/news/scientist-of-the-day/deodat-de-dolomieu/. Linda Hall Library, located at the University of Missouri—Kansas City, is an independent science research library with collections in science, engineering, and technology.

11. Daniel Beckman and Steve Jones, "Ecological Health of the Upper White River Basin: A Technical Report to the National Fish and Wildlife Foundation," James River Basin Partnership, 2002. Beckman and Jones taught biology at the university level and conducted numerous studies on the fish fauna of the Ozarks.

12. Dave Hamilton, "Missouri River Otter Saga," in, *Missouri Conservationist,* Vol. 68, Issue 6, June 2007. Hamilton, a long-time research scientist with the Missouri Department of Conservation, passed away in September 2007.

13. Jack H. Ray and Neal H. Lopinot, *Projectile Point Types in Missouri and Portions of Adjacent States,* Missouri Archaeological Society, 2016. Ray and Lopinot, with the Center for Archaeological Research at Missouri State University, have authored numerous papers based on their research at archaeological sites all over the Ozarks.

14. *Rude Pursuits and Rugged Peaks,* p. 78

15. Ibid., p. 79

16. Elmo Ingenthron, *The Land of Taney: A History of an Ozark Commonwealth,* Ozark Regional History Series Book II, Point Lookout, Mo., 1974. Ingenthron (1911-1988) was a long-time resident of Taney County, serving as a school principal at Branson, Blue Eye and Abesville. He was a charter member of the White River Historical Society and authored four books on the Missouri Ozarks.

17. *Rude Pursuits and Rugged Peaks,* p. 52

18. Ibid., p. 52

19. Ibid., p. 80

20. Ibid., p. 81

21. Carl O. Sauer, *The Geography of the Ozark Highland of Missouri,* University of Chicago Press, 1920, p. 63. Sauer (1889-1975) wrote about the geography of the Ozarks in his dissertation while studying at the University of Chicago. He later taught geology at the University of Michigan and University of California at Berkeley.

22. Vox website: https://www.vox.com/2015/8/7/9111123/whiskey-civil-war-chart
Vox is an American news and opinion website owned by Vox Media, founded in 2014.

23. *The Geography of the Ozark Highland of Missouri,* p. 178

24. Robert B. Jacobson and Alexander T. Primm, *Historical Land-Use Changes and Potential Effects on Stream Disturbance in the Missouri Ozark Plateaus, Missouri,* U.S. Geological Survey Open-File Report 94-333, Rolla, Mo., 1994. Jacobson, long time researcher with the Geological Survey, is now tackling "big river issues," such as restoring habitat for the endangered pallid sturgeon in the Missouri River. Primm, an oral historian, recently published *Ozark Voices: Oral Histories from the Heartland.*

25. Phyllis Rossiter, *A Living History of the Ozarks,* Pelican Publishing Co., Elmwood, La., 1992. Rossiter (1938-2007), longtime resident of Ozark County, Missouri, wrote many fiction and non-fiction books about the Ozarks. She had a regular column, "A Sense of Place," in *Ozark Mountaineer Magazine.*

26. United States Department of Agriculture, "Climate Hubs" website: https://www.climatehubs.usda.gov/hubs/southwest/topic/feral-swine-bomb. This website provides information and maps describing the historical spread of feral swine, along with photographs of the damage they do to crops and waterways.

27. Henry R. Schoolcraft, *A View of the Lead Mines of Missouri: Including Some Observations on the Mineralogy, Geology, Geography, Antiquities Soil, Climate, Population, and Productions of Missouri and Arkansas and Other Sections of the Western Country*, New York, 1819, p. 74. Schoolcraft wrote two books about his travels in the Missouri Ozarks in 1818-1819. *A View of the Lead Mines* was more technical, describing the methods of mining and smelting lead ore at Potosi. While Schoolcraft was examining the mines and conducting experiments on the practical uses of lead, he stayed at the mansion of Moses Austin, the famous Potosi lead miner. Schoolcraft's *Journal of a Tour,* published in London in 1821, was more of a travel adventure intended for lay audiences, describing his experiences with the people and places of the Missouri backwoods.

28. *A View of the Lead Mines,* p. 16

29. Arthur Winslow, *Lead and Zinc Deposits,* Missouri Geological Survey, Vol. 7, Rolla, 1894. Winslow served as Missouri's State Geologist from 1889 to 1894. In 1894, he was fired by the Survey's Board of Managers for failing to produce reports on time.

30. Missouri State Highway Commission, *Missouri Highways: The First 200 Years,* Missouri State Highway Commission, 1966

Index Topics

www.ingramcontent.com/pod-product-compliance
Lightning Source LLC
LaVergne TN
LVHW082352200926
848481LV00012B/3539
* 9 7 9 8 9 9 4 7 7 2 1 2 6 *